The Why of Things

The Why of Things

Causality in Science, Medicine, and Life

Peter V. Rabins

Columbia University Press

New York

Columbia University Press
Publishers Since 1893
New York   Chichester, West Sussex
cup.columbia.edu
Copyright © 2013 Peter V. Rabins
All rights reserved

Library of Congress Cataloging-in-Publication Data
Rabins, Peter V.
The why of things : causality in science, medicine, and life /
Peter V. Rabins.
pages cm
Includes bibliographical references and index.
ISBN 978-0-231-16472-6 (cloth : alk. paper)
—ISBN 978-0-231-53545-8 (e-book)
I. Causation. 2. Science—Philosophy. 3. Life sciences—Philosophy.
I. Title.

Q175.32.C38R33   2013
122—dc23   2012051072

Columbia University Press books are printed on permanent
and durable acid-free paper.
This book is printed on paper with recycled content.
Printed in the United States of America

c 10 9 8 7 6 5 4 3 2 1

Jacket design: Marc J. Cohen

References to websites (URLs) were accurate at the time of writing.
Neither the author nor Columbia University Press is responsible
for URLs that may have expired or changed since the manuscript
was prepared.

TO MY MENTORS
Donald Gallant
Paul McHugh
Philip Slavney
Marshal Folstein

Contents

Preface

The ideas developed in this book were first presented at a clinical teaching presentation to the faculty and students of the Department of Psychiatry and Behavioral Sciences at the Johns Hopkins School of Medicine. In this forum, the lecturer begins by discussing a clinical encounter with a patient and uses the issues raised by that person's situation to address a broader question. In trying to address my patient's question, "Why did this happen to me?" I realized that the same question arises for all of us in the course of work and personal life. That a topic such as causality would be considered an appropriate accompaniment to a clinical teaching presentation is a testament to the vision of the then department chair, Paul McHugh, that the practice of medicine should be rooted in an intellectually defensible and discussable framework. While the rough outline of the schema presented in that 1995 lecture is maintained here, the ideas have evolved in response to the input, questions, and criticisms provided by many colleagues, friends, and family members. To them I owe deep gratitude.

I have had many wonderful and influential teachers dating back to elementary school, and I have no doubt that the ideas presented here are an outgrowth of their teachings. The four individuals to whom this book is dedicated played a special role in my development as a physician and psychiatrist and shaped how I have approached both the scientific and the clinical work I have done throughout my career. Donald Gallant showed me that psychiatry could be intellectually rigorous, that many patients could be helped, and that bringing care to the places where the most disadvantaged lived would make a difference in the lives of many people. From Paul McHugh and Philip Slavney I learned many things, especially the importance of identifying one's core assumptions, modes of logic, and intellectual predecessors. Marshal Folstein guided my immersion in the interface between psychiatry and the brain and impressed upon me the importance of hypothesis testing.

I began writing this book during a three-month sabbatical in 2001. Its decade-long gestation reflects both the evolution of the ideas and the distillation that comes with many rewrites. The Johns Hopkins Berman Bioethics Institute provided office space during my sabbatical and a forum for presenting these ideas, and to its members I am grateful. Much of the writing and editing took place in the Plum Lake cabin of Marilyn and Peter Julius. Having this extraordinary place away from the distractions of clinical work, teaching, and administration helped me refine my thinking.

Philip Slavney gave a close reading to the first complete draft of the book, and his extensive input improved many aspects of my logic and presentation. My editor, Patrick Fitzgerald of Columbia University Press, was both supportive and critical and helped further improve the writing. The three anonymous reviewers he recruited to vet the typescript made many valuable suggestions, and to them I express gratitude.

My extraordinary family has been supportive throughout the writing of this book. Discussions with them have helped shape my ideas, and they have contributed to the artwork. They continue to inspire and teach me. My wife, Karen, read the final manuscript and, as she has in many of my writings, made significant contributions to the ideas and the writing.

The Why of Things

INTRODUCTION

Men are never satisfied until they know the "why" of a thing.

—Aristotle

On March 11, 2011, a tsunami struck the Fukushima Daiichi power plant on the northeastern coast of Japan. The plant had shut down, as planned, forty minutes previously, when an earthquake occurred just miles off the coast, but the tsunami destroyed the backup sources of electricity that powered the required constant cooling of the reactors. The resultant core meltdown of three of the facilities' five reactors led to a major release of radiation.

What caused this catastrophic failure? The most straightforward answer is the earthquake and tsunami. But subsequent expert analyses cited "technical and institutional weaknesses," such as a weak authority structure at the plant and within the company that managed it and the voluntary nature of the standards by which nuclear power plants are managed and overseen. Still others pointed to the plant designers' failure to provide a mechanism by which cooling could continue in the face of prolonged power loss and their decision to build so many reactors at a single site.

Thirty-two years earlier, in March 1979, the nuclear power plant at Three Mile Island, Pennsylvania, experienced a catastrophic failure. The precipitating event was an open valve that triggered a series of events ending in reactor failure. In his book on the catastrophe, entitled *Normal Accidents*, the sociologist Charles Perrow concluded that the complexity of modern industrial production facilities, especially nuclear power plants, makes catastrophic failure inevitable and predicted such an accident every decade (Chernobyl and Fukushima have followed in the next thirty-two years). Perrow identified a number of causes that contributed to the Three Mile Island failure, including the multiplicity of interacting elements at the plant and an unwillingness by the many groups of people involved in design, management, political approval, financing, and disaster preparedness to accept the inability of humans to anticipate all potential sources of failure—an attitude best characterized as human hubris.

In my work as a psychiatrist over the past thirty-five years, I have often been asked questions about cause: "Why have I become depressed? Is it something I did or should have done? Or is it some experience of mine in the past?" "Is it genetic since my mother was treated for depression?" "Is this a punishment from God?" "Why do I seem to become friends with people who ultimately turn on me?" "Why do I repeatedly get into trouble with my bosses and lose jobs?"

It is questions such as these that led to the writing of this book. These "why" questions seem so natural to ask, and so important, that many people are convinced that they should be answerable. Yet the answers to questions such as why the Fukushima and Three Mile Island disasters occurred or why a person becomes depressed are complex and multifactorial. How can we include factors as disparate as a valve left open, the inherent complexity of multisystem manufacturing plants, and the inability of humans to anticipate all of the potential errors and adverse events in operating such a complex system? How can genetics, early life experience, and current events be understood as causing depression in one person but not in another with similar experiences and background? How can one choose where to begin? What are the rules or standards by which answers should be judged? Is there even a standard? Is the task impossible because there is no way to judge a correct answer?

The solution proposed here is a pluralistic approach that assumes that there is a best approach for each question and that it is the job of the seeker to determine which method or combination of approaches is best suited to the question being asked. This book proposes a three-facet model of causality. As a preview:

- Facet 1 consists of three *conceptual models* of causal logic: the unclosed valve in Three Mile Island is an example of the yes/no or *categorical* model. The genetic contribution to developing depression is likely a graded, *probabilistic* risk rather than an absolute yes/no. A depression that occurs after a relatively minor stress that followed a long string of moderate or severe stressors would be an example of an *emergent* or nonlinear cause.

- Facet 2 describes four *levels of analysis*, an approach first suggested by Aristotle 2,400 years ago. In the Three Mile Island and Fukushima examples, *predisposing* causes were the flawed training and management oversight; the tsunami was a *precipitating* cause. The inherent complexity of the many interacting systems that make up a nuclear power plant is a *programmatic* cause, and human hubris is a *purposive* cause.

- Facet 3 describes the three *logics* by which knowledge of cause is gained. The *empirical* method uses the scientific method, for example, the determination that a genetic variant is present in multiple members of a family in which depression is common. The *empathic* method uses the logic of narrative connectedness to support the reasoning that a specific stressor is negative for one person but not another. *Ecclesiastic* logic would be employed by a believer who attributes cause to an actual lapse in his longstanding participation in the precepts of his religion.

A helpful way to visualize these three facets is as the three sides of a tetrahedron, as shown in the figure here and the diagrams that open each chapter. These diagrams reinforce several important aspects of this approach. First, the three facets are not totally separate but can (and should) be used in combination when appropriate. Second, they are not hierarchical. To help the reader along, each chapter opens with an image of the facet or facets that it will focus on.

The proposed three-facet model is complex, even daunting, and the burden is on the book to justify this complicated approach. I have come to it because several broad challenges must be addressed

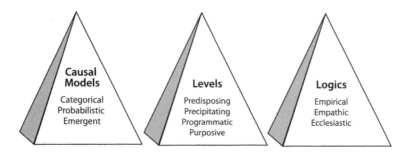

by anyone interested in exploring how causal attribution can be justified. First, there is no single definition of cause. Second, the understanding of cause has varied over time and across cultures. And third, one cannot "prove" the existence of the concept of cause or causality. As a result, this book is built on the *premise* that *causes exist* and that *causal relationships can be discovered and confirmed*. This must be stated as an assumption because it is not only impossible to prove but also, as will be discussed later in the book, impossible to disprove. In fact, some well-respected scholars and thinkers claim that the concept of cause is nothing more than a convenience that has no meaning besides its uses in everyday discourse, applied science, logic, or religion. Others cite the widely varying view of causality across cultures as evidence that it is a convention rather than a valid or universal construct.

To make the issue even more complex, there is no single definition or method for determining "the why of a thing." How can it be that there is no one right or best method for determining cause but that causes exist and can be accurately identified and that a best way of approaching a specific question of cause is sometimes possible? I believe an answer to this question emerges when one appreciates that the definition of cause, the history of the concept's development, and the establishment of methods for determining cause are all intertwined. The pluralistic model presented here is built upon an amalgamation of these three big questions. Examples will be used throughout the book to illustrate the types of questions that a particular method or model is useful in addressing. These examples will also illustrate both the strengths and limitations of that approach and should make it clear that the attempt to identify general principles often oversimplifies what happens in the real world. The reader is encouraged to bring a healthy skepticism to these discussions and to use the examples as checks on whether the arguments that are being put forth have validity.

1

HISTORICAL OVERVIEW
The Four Approaches to Causality

A thing cannot occur without a cause that produces it.

—Pierre Simon de Laplace

The concept of causality is so much a part of our lives that we often think about, discuss, or identify causes without considering the complexity of the underlying concept. Questions about cause touch on issues small and large—questions such as, why did I stumble, what led to that car accident, what caused today's weather? Why are some people happier than others? Why do some individuals become sick while others avoid an illness that "everyone is getting"? What causes poverty, economic cycles, substance abuse, evil? How did the universe come to be?

It is not the goal of this book to answer any of these questions with absolute certainty—that would not be possible. Rather, its goal is to provide *approaches* to answering such questions. We will begin by trying to understand what we mean by the word "cause," since understanding what lies behind the words we use can help focus a search, clarify what is being sought, and settle some of the arguments that arise. This approach shares an assumption, one dating from the ancient Greeks, that human reasoning can be a source of knowledge.

Understanding what we mean by cause is a big question. It is the "why" question asked by two-year-old children, mature adults, historians, geneticists, clergy, and ethicists. Many great thinkers in disciplines as disparate as theology, philosophy, neurophysiology, history, particle physics, and accident prevention have given thought to what is meant by "a cause." To begin to answer the question, "What do we mean by causality?" we will survey some of the major ideas. This will give the reader an appreciation of how current concepts have developed over time and identify some of the major challenges faced by anyone interested in the question.

However, taking a historical approach in this instance does not mean that ideas and concepts emerged in a specific, ordered sequence or developed in a progressive, linear path. Many of the concepts discussed here developed in widely separated places and reached other parts of the world only centuries later. This is clearly illustrated in the development of similar aspects of causality that emerged in the Eastern and Western worlds at different times and without apparent influence from the other sphere. The benefit of the sequential approach is that it provides a structure upon which disparate ideas can be hung and shows how concepts of causality have changed over time. Perhaps even more importantly, tracing the development of the concept of causality over human written history emphasizes the long struggle that humans have had with the issue and supports the notion that a complete understanding may never develop. A historical review also reveals that our current concepts of causality are an amalgam of ideas that have arisen and developed over thousands of years. They reflect on and derive from long traditions of thought that have engaged many groups and cultures. By necessity, this review will be selective. It will highlight some of the most challenging and contentious issues and set the stage for much of the discussion in succeeding chapters.

HISTORICAL OVERVIEW

The earliest human writings demonstrate the centrality of the concept of causality to humankind. Five-thousand-year-old Sumerian cuneiform tablets and 3,500-year-old Egyptian papyri identify forces or beings that brought about (caused) the world in the way that these cultures (or at least the authors) experienced it. Ancient religious tracts such as the Hindu Rig Veda and the Aramaic Tanakh or Old Testament do the same. These ancient texts also link events of the present to the influence of the initiating being or force. The identification of an initiating cause as the explanation for the universe as we know it not only persists today in many religions but also is a central tenet of such scientific hypotheses as the Big Bang theory, which ascribes the makeup of the current physical universe to events that occurred at the instant of its formation, and the pantheistic Gaia concept, which describes the Earth as an organism constructed such that a change in one aspect leads to an adjustment in others in order to sustain equilibrium. Thus, what is today labeled a supernatural origin of events can be found in many if not all cultures and must be addressed if a thorough accounting of the concept of causality is attempted. This will be the focus of chapter 10.

The idea that individual humans can cause events has also been present in Eastern and Western thought for thousands of years, although it is not possible to prove that all groups of human beings have conceptualized causality in this fashion. The Hindu concept of karma, which assigns to individuals responsibility for their actions and explains the form into which a person is reincarnated as a result of past choices, implies that individuals are the agents of cause. The oldest extant compilation of laws, the Code of Hammurabi, which dates to 1750 BCE, likewise assigns to individuals the responsibility for their actions, as do the biblical stories of Adam and Eve, the Flood, and the Ten Commandments. The importance of these documents demonstrates that the concept of causal agency has long been a central aspect of human thought.

Several hundred years after the story of Moses is said to have occurred, the Greeks developed the Western tradition of analytic thinking as a source of knowledge. Democritus (c. 400 BCE) conceived of events as having ultimate single causes, although he suggested that causality could be so complex that it was often hidden from human observation or at the least very difficult to discern. At about the same time, Plato proposed that objects like chairs and concepts such as cause exist as ideals against which actual chairs and causes can be measured or compared.

Plato's idea that we use ideal models or "exemplars" as a standard against which an actual event is measured has been shown in recent cognitive neuroscience experiments to be an innate human approach. Inherent in it are two theses that will recur throughout this book. First is the idea that once a standard is identified, it can be approached closer and closer over time, even though perfection is never achieved. Second is the idea that the ideal exists as much in the abstract as in actuality. Plato never applied these concepts to the study of cause. Nevertheless, they underpin the approach taken in this book, which holds that it is possible to develop a model of causality that gets closer to the ideal over time by incorporating into the concept those ideas that improve its accuracy and jettisoning those that are no longer helpful. The implication that the Platonic approach has resulted in an increasingly nuanced and deeper concept of causality is embraced by this book, and so is the recognition that a complete and permanent definition cannot be achieved.

Plato's pupil Aristotle proposed a multifactorial model of cause and effect that describes cause as existing at several different levels of analysis. Table 1.1 lists the four levels of causality that he identified, provides my adaptation of them, and provides a commonly cited example from Aristotle's writings. Aristotle's meaning of "cause" was different than what is generally meant in the present era, but his conception is still strikingly modern. He describes the "cause" of a

Table 1.1   Aristotelian Model of Cause

Aristotelian Term	New Descriptive Term	Definition	Bronze Statue of Zeus
Material	Predisposing	Inherent, preexisting	Strength and malleability of bronze
Efficient	Precipitating	Initiating, provoking	The sculptor
Formal	Programmatic	Systemic, interactional	The beauty of the ideal human body
Final	Purposive	Reason, teleology	To inspire and honor

statue. The *material cause* is the bronze and the unique properties that make that alloy desirable for the production of a representation of a detailed human body. The *formal cause* is the conception of the ideal body and the concept of making an idealized representation. The *efficient cause* is the artisan and the skill the artisan brings to the process. The *final* cause is the purpose of the statue, for example, to exalt the ideal human body or to honor the god represented. "What causes a statue?" is clearly a question about what brought it into being, a question that addresses one aspect of causality, but it is not a question of primary interest today. Although complex, Aristotle's multifactor, multilevel model was extraordinarily influential for almost two thousand years. For example, when Thomas Aquinas (c. 1225–1274) discussed cause in a theological context, he conceptualized God as operating at each Aristotelian level. Chapter 2 will present an expanded and modified version of the Aristotelian model adapted to current questions.

With the emergence and development of the scientific method, the conceptualization of cause and the methods for demonstrating causality have undergone major changes over the past four hundred years. Although the scientific method as we know it today had no single beginning, Francis Bacon (1561–1626), in his *Novum Organum* (1620), is often cited as the first individual to recognize its characteristics and potential. Not an experimentalist himself, Bacon nonetheless recognized that an approach to knowledge that combined the three elements of repeated observation, integration of positive (confirming) and negative (disconfirming) results, and skepticism toward authority as the primary source of accurate information signaled a new way of seeking knowledge. He cited the Danish astronomer Tycho Brahe's voluminous collection of data on the movement of celestial bodies and subsequent 1512 discovery of a nova (which demonstrated that the universe was not static, contradicting a basic Aristotelian precept) as examples of this new approach to knowledge acquisition.

Other discoveries in the sixteenth century further contradicted the Aristotelian model of the universe and undermined the absolute acceptance of Aristotelian intellectual authority. For example, Copernicus's claim that the sun, not the Earth, was at the center of the solar system (his book *De Revolutionibus* was published at the time of his death in 1543) was supported by Johannes Kepler's (1571–1630) demonstration that the planets' motion could be described mathematically as ellipses, not perfect circles as Aristotle claimed, and Galileo Galilei's (1564–1642) identification of moons revolving around Jupiter contradicted Aristotle's claim that celestial bodies revolved only around the Earth. Furthermore, his belief that moving objects naturally slow down was replaced by Galileo's demonstration that falling bodies accelerate at a uniform rate and by Isaac Newton's (1642–1727) concept of momentum—enshrined as his first law of motion—that objects continue to move in the same direction and at the same velocity unless acted upon by an external force.

Galileo directly attacked the Aristotelian model of cause in his book *Discourse on Two New Sciences* (1638). He proposed that new knowledge is best gained by observation and measurement, not introspection. In the *Discourse*, the character representing Galileo's point of view cites his ability to describe the acceleration of falling bodies mathematically but his concomitant inability to identify the cause of the acceleration as evidence that the search for an Aristotelian final cause is futile.

Galileo's rejection of the Aristotelian idea that cause has multiple meanings and his emphasis on identifying questions or events in which direct measurements can be made (similar to the aspect of cause that Aristotle referred to as "efficient") established a narrowed concept of cause that persists today. I will refer to this narrower definition of cause as the *categorical model* because it seeks as causes single events that are either present or absent. This model will be discussed in detail in chapter 3. As noted earlier, this narrowed concept of cause predated Aristotle, but the Aristotelian model so overshadowed it that the categorical approach only regained a prominent role with the emergence of the scientific method in the seventeenth century.

Another of Galileo's ideas that has been influential in scholarship about causality was highlighted by John Stuart Mill two hundred years later in his use of the phrase "necessary and sufficient." This conception of cause states that A is a necessary and sufficient cause of B *if* A always occurs before B *and* B never occurs without A. This is a very high standard: it implies that an event can have only one cause. This standard cannot be applied in many situations. However, when it does describe a situation, the likelihood of a causal relationship is high.

While the Galilean view has been presented here as a radical move away from the Aristotelian multifactor, multilevel model of causality, this view becomes absolute only in retrospect. Even the scientists of the time had no sense that the pre-Galilean conceptions of causality had been overthrown. For example, both Isaac Newton and Gottfried Wilhelm von Leibniz, two of the most accomplished and best-known scientists (and sometime rivals) of the seventeenth century, wrote philosophical tracts that identified God as the ultimate cause, much as had Thomas Aquinas four centuries earlier. Newton believed that the regularity of the laws that he discovered demonstrated that they were manifestations of God's work, and Leibniz believed that the organization of the world reflected God's plan and was, therefore, the best possible manner in which the world could be organized. Both Leibniz and Newton saw a role for experimental and mathematical study but remained convinced that God was the ultimate explanation. Neither saw this dual model as a contradiction but rather conceptualized science and religion as complementary causal models that confirmed each other.

While the Galilean criticism of Aristotle might be characterized as a reemphasis on precipitating cause rather than a rejection of the Aristotelian model, the change was a radical one, and it significantly influenced the approach to cause over the next 350 years. It made the search for "sufficient" elements the defining criterion of causality and narrowed the search for causes to observable and testable elements. In effect, it defined the essence of causality as the identification of *precipitating* events. What accounted for—what caused—this dramatic development? I suggest it was the concatenation of events in the West during the sixteenth and seventeenth centuries. They included new technologies, such as the telescope; new methods for gathering and analyzing information, such as probabilistic models; great thinkers such as Leibniz, Newton, and Galileo; economic changes that provided leisure time and fiscal support for brilliant individuals to pursue new knowledge outside of the Church; the development of the printing press, which provided a method of broad and relatively rapid information transfer; and the emergence throughout Europe of educational institutions in which individuals who could put together the new observations, technologies, and methods worked. (This is an example of narrative logic, discussed further in chapter 9.)

LIMITATIONS OF THE SCIENTIFIC METHOD

Doubts about the scientific method's ability to identify causes quickly followed, however, even among individuals who were practitioners of science. For example, René Descartes (1596–1650), an experimentalist whose contributions included Cartesian two-coordinate geometry, the idea that mathematical relationships underlie the basis of physics, and the concept of momentum, expressed skepticism about the ability to gain knowledge through observation alone. He proposed that one should start from stated principles and *deduce* truths from them. This led to his claim that one could begin from the statement *Cogito ergo sum* (I am thinking, therefore I exist) and deduce both the existence of God and the duality of the mind and the body.

Such skepticism about relying upon the senses can be traced back to the Greek Stoics fifteen hundred years earlier, but it is Descartes's suggestion that the method of deduction is the most useful method for identifying causes that deserves attention here because the deductive method is still with us and because Descartes's applications of it demonstrate that what one claims to deduce is still open to challenge.

Galileo's and Bacon's renewed emphasis on precipitation as the defining feature of causality also came under challenge from the Scottish philosopher David Hume (1711–1776). Hume claimed that causality could never be definitively demonstrated because it relied upon *inductive* reasoning, that is, it required a leap of belief that two events were inevitably linked and, thus, the drawing of conclusions that go beyond the facts. Even if an event B always follows event A, Hume argued, one could only "guess" that A had caused B. Such associations could never *prove* causality.

Hume did not totally dismiss induction, however, but said it could never establish causality with certainty. Hume's skepticism about inductive reasoning persists today, both among scientists who object to seeking broad explanations for natural phenomena and among antiscientists who reject the scientific method as a means of increasing knowledge and understanding.

As pointed out by Karl Popper two hundred years later, Hume's rejection of induction is itself an induction. Nevertheless, Hume identified an important caveat: inductive reasoning has unavoidable limitations and cannot absolutely "prove" that two events are causally related. It is equally important to emphasize, though, that Hume did not claim that the search for cause is futile. He cited the repeated demonstration that two events occur together and the identification of multiple lines of evidence that point in the same direction as support but not proof of causal relationships.

Even today, though the caveat that the occurrence of two events together (association) does not indicate causality is widely recognized, Hume's identification of this limitation in the search for causal relationships is often ignored. The seduction of inductive reasoning is a trap that easily catches the unwary. Being thoughtful about the meaning of "cause," using caution when claiming such a relationship exists, and requiring multiple lines of evidence can lessen the chance that one will be wrong. Hume's skepticism challenged humankind's readiness to accept causality as a given, but it spurred a refinement of the concept and underlies much of Western thought about the subject during the subsequent two hundred years.

At the same time that Hume was expressing skepticism about the possibility of identifying cause with absolute certainty, the Italian philosopher Giambattista Vico (1668–1744) was expressing similar concerns about the validity of causal knowledge in the discipline of history. Vico noted that most proposed causal mechanisms found in historical writing derived from an analysis of events *after* they had happened. He suggested that a primary distinction should be made between information gathered by scientific and nonscientific methods. His concern will be examined in depth in chapter 9.

IMMANUEL KANT AND THE ROLE OF
HUMAN PERCEPTION

Hume's radical dismissal of induction spurred Immanuel Kant (1724–1804) to reformulate the concepts of cause and causality less than half a century later. Kant proposed that humans impose upon nature basic categories such as causation. Stated in more modern terms, this proposal states that the organization of the nervous system determines the way in which things are perceived. Kant extended this hypothesis to the issue of causality and proposed that the concept of cause is an innate aspect of human thought. Thus, causes exist because the human brain is organized to conceive of causal relationships among events. This extraordinarily radical idea (although it did have precedents among the ancient Greeks) has received support from several lines of modern experiments. Patients who have undergone "split" brain surgery, for example, can be shown to experience and think about causal relationships linking two events differently in each of the disconnected halves of their brain. Research with infants also suggests that the notion of a causal relationship develops between years two and three, but the interpretation of these experiments depends upon agreement that certain behaviors indicate the presence of the concept of causality, an interpretation that is not universally shared. Recent MRI scan studies also suggest a neural basis for human categorization.

MEDICINE OPERATIONALIZES CAUSE

By the latter half of the nineteenth century, new technologies and intellectual approaches built upon the strengths of the "direct-agent" model of cause advocated by Galileo and his successors. This is well illustrated by advances in medical knowledge. For thousands of years, physicians who wrote about medicine focused on individual symptoms such as fever, shortness of breath, seizure, and confusion. Each was considered a specific entity, much as we today consider individual diseases to be distinct conditions. However, in the mid-seventeenth century, the British physician Thomas Sydenham (1624–1689) observed that certain medical symptoms clustered together with regularity in many patients. He suggested that these groupings of medical symptoms, now called *syndromes*, represented actual entities and proposed a test of this hypothesis: they would be found in patients from different parts of the world and in different historical epochs. Furthermore, he proposed, the existence of these entities was proven by the fact that each would follow a predictable course over time and have a predictable outcome regardless of where or when the person lived. For example, patients who presented with the three symptoms of fever, cough, and sputum production would likely be suffering from pneumonia, a disease of the lungs, while patients with fever, stiff neck, and confusion were likely to be suffering from meningitis, a disease of the lining of the brain and central nervous system.

This dramatic new approach introduced the concept of disease as we understand it today. Two hundred years later, this concept was linked with the autopsy to develop a method called the clinical/pathological correlation by such nineteenth-century physicians such as the German pathologist Rudolf Virchow (1821–1902). This linkage provided a means of demonstrating that many patients with the syndromes identified by Sydenham's method of clustering symptoms had the same bodily abnormalities at autopsy and thus provided a method for demonstrating that a specific bodily abnormality was causative of a specific disease.

More relevant to this discussion, the clinical/pathological approach became a method by which the cause of a specific disease could be established. For example, linking specific abnormalities in bodily structures to specific clinical syndromes led to the abandonment of the beliefs, dating back to the ancient Greeks, that sickness was caused by imbalances of bodily humors (black bile, yellow bile, phlegm, and blood) and environmental substances ("miasmas"). What the autopsy offered was a way to "prove" such linkages and thus to prove the value and specificity of the model proposed by Sydenham. The modern conception of disease that derived from this model is still broadly accepted both within the profession of medicine and by the public. Many of the advances that have occurred in medicine over the past hundred years attest to the strengths of this model of causality, but, as we will see in chapters 4, 5, and 8, some of its failures also derive from the limitations of an overly simple model of disease causality.

Another great medical discovery of the nineteenth century, the germ theory, led directly to a codification of methods for establishing causality in experimental and clinical medicine. In experiments carried out in the middle of the nineteenth century, Louis Pasteur (1822–

1895) and others found that microscopic organisms called bacteria were found to be associated with many syndromes such as pneumonia and meningitis. But how could these causal links be proven?

The microbiologist Robert Koch (1843–1910) proposed three criteria, later termed Koch's postulates, to prove that an organism caused an infection:

1. The organism is repeatedly isolated from individuals with a specific disease;
2. the organism is then reproduced in such a quantity that
3. upon introduction into either animals or human beings, the initial disease is replicated.

This schema contains elements of Hume's suggestion that repeated association strengthens the likelihood of a causal association and of the Galilean idea that causality implies that the relationship between the two events is *necessary*, that is, that the disease would not occur without the agent. The criteria have been modified over the past century and now include an element of *sufficiency*, that is, the idea that the disease never occurs without the agent or that the disease ceases to exist if the agent is removed, for example, by treatment. These criteria describe the essence of the scientific process by which a cause or causes is identified in a biological system. It is a powerful application of the single-cause disease model that will be discussed in chapter 3. Stated more generally, these criteria postulate that A can be demonstrated to be a cause of B if

1. A is *repeatedly* associated with B (correlated or associated with); *and*
2. B occurs regularly when A is introduced (sufficient); *and*
3. The removal of A leads to a resolution of B (necessary).

However, Koch's postulates or causal criteria do not explain several issues in causality that relate to the arena of microbiology and to causality more broadly. Why do some individuals who are inoculated with an organism not develop the disease? Why does the same strain of organism cause variable manifestations in different individuals?

Why do diseases vary in their frequency or prevalence in different geographic areas? These questions reveal that there are limits to the universality of the postulates, but the advances in knowledge that have resulted from their application over the past hundred years are a testament to their power and utility. The rapid linkage of the HIV organism to the immunodeficiency syndrome AIDS, for example, used just such logic, although criterion 3 was not demonstrated in humans for ten years after the virus's discovery.

The questions in the last paragraph that are not answered by Koch's postulates illustrate a much broader issue: *our ability to come up with general rules for establishing causality will always be limited by the specifics of a given causal question.* In the example of proving that a specific infectious agent is the cause of a specific disease, there is variability not only among the organisms known to cause a specific illness (for example, some may have a gene that confers antibiotic resistance while others do not) but also among the individuals who are infected ("host" immune factors) and differences among the environments in which the host and agent are residing. In this example, then, there are three elements in the causal chain, the agent, the host, and the environment, and aspects of all three influence the event of interest (here an infectious illness) and its causal chain. This issue will be encountered in a number of guises throughout this book. It is both frequent and important enough to state as a general statement:

> The ability to predict cause in a single encounter is influenced not only by specifics of the potential causative agent A and specifics of the object O being acted upon but also by specifics of the environment E in which they occur.

This limitation in the ability to determine causality echoes Hume's identification of the limits of induction. Every replication of event A is not an exact copy of that event—each instance is unique no matter how carefully the situation is manipulated to make it the same. This limit to replication identifies a limitation of our ability to generalize about cause, but a number of steps can be taken in the experimental situation to minimize greatly any differences. The many successes of microbiology and the successful application in many disciplines of the reasoning encapsulated in Koch's postulates demonstrate that accurate generalizations can be made. Thus, general rules about causality can be derived, but there will likely always be exceptions to them, and therefore they will likely always have limitations. (Of course, this statement is itself a generalization, so if it has a limitation then the result would be an exception to the exception—that is, the claim that a generalization can exist that does not have limitations. This reflects a limitation of the rhetorical method used in empathic reasoning [see chapter 9].)

Identifying the limitations of any set of criteria for identifying causal relationships is an important step in establishing the existence of such guidelines. The identification of limitations actually strengthens the search for causal relationships; by emphasizing that causal relationships cannot be established with absolute certainty, the recognition of the limitations of any approach should spur the users to identify other information that further corroborates a proposed relationship. As will be discussed in chapter 4, methods for estimating the likelihood of a relationship have been developed over the past several hundred years, and it is possible to state that the likelihood of a relationship is close to or almost absolute. The word "almost" will be rejected by some who believe that is it possible to establish causality with absolute certainty and will lead others to conclude that it is impossible to establish causal relationships. What is being proposed here is a limitation to the concept of causality, a limitation that ultimately strengthens the concept through embracing these limitations:

> No single set of rules can be established by which causality can be proven. Because every event is unique in time, there will always be specific elements of a situation that influence the outcome but cannot be enumerated. Nevertheless, causality can be established with quantifiable certainty even if absolute certainty is not possible.

THE IDEA THAT CAUSE CAN BE ANALYZED REEMERGES

By the end of the nineteenth century, the scientific method had become widely accepted as the approach for establishing causal mechanisms in disciplines as diverse as medicine, physics, and biology, and many conceived of it as the single method by which causality could be proven. Hume's critique of the inductive method was well known among philosophers but ignored or dismissed by many in the sciences. However, the difficulty of applying a similar standard to the field of history, first pointed out by Vico a half century before Hume, began to be appreciated and led many to distinguish between the "true" sciences and the "social" sciences.

Max Weber (1864–1920), a founder of the discipline of sociology, sought to resolve this tension between science and history by proposing that there are two approaches to establishing causality, one linked to the study of scientific issues and the other appropriate to the study of history. This proposal was similar to Vico's suggestion of two centuries earlier. This idea was further developed by Karl Jaspers (1883–1969), who trained in psychiatry and later became an eminent existential philosopher, in his book *General Psychopathology*, first published in 1913.

Jaspers described two conceptual models of cause. He called one *verklaren*, the method of *causal explanation*. This approach determines causality in situations in which phenomena can be observed by multiple observers and, ideally, can be replicated on several occasions. Galileo's experiments, Koch's postulates, and the clinical-pathologic method are examples of disciplines in which *verklaren* is appropriately employed. Jaspers labeled the second approach to establishing linkages among events *verstehen*, the logic of *meaningful understanding*. *Verstehen* relies upon the intuitive appreciation of a situation and the linkages among events that led to it. Accuracy in *verstehen* is established when a consensus develops among knowledgeable people that a particular causal linkage is accurate.

Weber and Jaspers believed that these two types of causal logic complement each other. They conceptualized them as distinct methods for determining two different types of causality and believed that their proper use is determined by the situation being considered. As a result, they did not view them as conflicting or competitive but rather as complementary approaches that are appropriately employed in different situations. It is not surprising, therefore, that different criteria are used to determine their accuracy. According to this Weberian model, the skill in employing these models of causality depends upon knowing when it is appropriate to apply each method.

The idea that there is more than one model of causality wasn't new, of course. The Aristotelian model had proposed the use of multiple approaches to the understanding of causal relationships more than 2,400 years previous. What set the Weberian model apart was its suggestion that the different models should be applied to different circumstances, whereas the Aristotelian model applies different levels of analysis to the same issue. Like the Aristotelian approach, the Weberian model uses the question being asked as the guide for choosing which method of analysis is best. It does not solve the challenge raised by Hume, since it does not remove the need for an inductive leap. However, it does clarify that the basis of the inductive leap might well be different with different questions. In the social sciences, the "leap" of inductive reasoning relies on an empathic understanding that two events are linked in a causal fashion. In the physical and biological sciences, this leap requires no such linkage, although, as will be discussed in chapters 6, 7, and 8, plausibility is sometimes a criterion in the sciences as well.

It seems safe to say that Weber's proposal has not been widely appreciated or accepted. Not only do most people seem to believe that there is only one approach to causality, but even individuals who would accept the proposal that more than one method or approach exists tend believe that one method is better than the other. Indeed, such disagreements are at the heart of many current debates about causality. One result is that each of the two models proposed by Weber has strong adherents who diminish the approach of the other with such descriptions as "unscientific," "unprovable," or "cold and unfeeling." An alternate reason for dismissing the distinction is accepting Hume's emphasis on the leap of judgment being made rather than whether the question being addressed is a "physical" or a "social" science question. This book will accept the Weberian proposal that different types of reasoning are useful for addressing different types of causal questions in spite of the recognition that the distinctions are not absolute. The clarity of the basis and criteria for making the distinctions is a challenge that still must be addressed, however.

The distinction proposed by Weber is not merely the science/nonscience divide described by C. P. Snow in his influential book *Two Cultures*. Many historians, political scientists, and sociologists utilize information such as birth rates and paper documents to determine plausible causal relationships among events and draw their conclusions from what almost everyone would conclude are data. Conversely, many biological and physical scientists speculate about the causes of single events such as the origin of the universe or why natural selection would choose one path over another or extrapolate from the behavior of mice and rats to human behavior and draw conclusions about violence, bonding, or depression. The widespread use of both *verstehen* and *verklaren* by individuals and groups who see themselves as using only one of these methods is an error in reasoning. Individuals who see themselves as scientists use both, as do practitioners of the humanities and social sciences. Hume's great contribution was to demonstrate that every experiment requires the leap of judgment of *verklaren*. Conversely, many applications of *verstehen* reasoning depend on information (data) gathered through the methods of observation that can be examined at least for reliability.

Another example of a debate that rests upon slavish adherence to a single method of analysis is the long-standing conflict between causal claims in science and religion. This conflict may have led to the demise of Islamic science in the eleventh and twelfth centuries and to the development of science as a distinct discipline in Europe soon after Galileo's forced recantation of the ideas he presented in *The Two Sciences*. These tensions exist today in debates between certain religionists and some scientists over evolution and the origin of the universe. Individuals on each side of the debate use a single (but different) method to claim that they have solved a very challenging question. They do so without acknowledging that they are beginning with very different assumptions and using very different models of causal reasoning and without recognizing that the answer that one seeks will be determined by what one is trying to explain. The solution proposed by Aristotle is to accept that several different approaches to identifying causality exist and that the most appropriate method for a given question depends upon the question being asked and the relationships between the causal factors being identified.

The solution proposed by Vico and Weber rests upon the claim that different types of knowledge exist (also an Aristotelian idea) and that different methods are appropriate for gaining understanding or truth about them. What these proposals share is a rejection of the notion that a single model or method exists by which causal knowledge is attainable.

PROBABILISM OVERTAKES MODERN PHYSICS AND SCIENCE

As discoveries in the physical sciences and radical new ideas in the social sciences challenged long-held basic concepts, the concept of causality underwent dramatic changes during the twentieth century. These new conceptualizations were especially noteworthy in physics, as relativity theory and quantum mechanics overturned well-established conventions about the makeup of the physical world.

Relativity theory emerged from a "thought experiment" (or speculation, if you prefer) in which Albert Einstein (1879–1955) demonstrated that the motion of an observer has a direct influence on what that person is observing. This conclusion followed from the then recent demonstration that the speed of light has a finite value (186,000 miles per second). If this is the case, reasoned Einstein, then time would stop for an observer moving at the speed of light away from a stationary event and would slow down for an observer moving slightly less than the speed of light. Events would seem to occur in the reverse order, so-called time travel, if that observer were moving faster than the speed of light. This idea contradicted the human experience that a sequence of events occurs in only one fashion and implies that the accuracy of an observation depends both upon the motion of observer and the motion of the observed event.

At a philosophical level (admittedly a big leap), relativity theory undermined the belief in absolute measurement and, by extension, the concepts of absolute accuracy. The widespread interest in relativity theory is demonstrated by the front-page placement of an article in the New York Times reporting that the 1919 eclipse of the sun corroborated the theory's prediction that light is bent by an object as massive as the sun.

One astounding implication of relativity theory is that time has no inherent direction—it can move in a "forward" or "backward" direction, depending on the velocity of the observer and the observed. This presents a major challenge to the model of cause that we have developed thus far, since we have proposed that the sequence in which events occur has a primary role in determining if there is a causal relationship. Indeed, at least since Aristotle, time had been considered to be a constant feature of nature. Although the philosopher Arthur Schopenhauer (1788–1860) had proposed that time is a human construct with no grounding in nature, Einstein's proposal that time could vary with the velocity of the observer provided a quantifiable and therefore testable statement of this relationship.

Clearly, this has significant implications for causality, because if directionality is not absolute, we must either abandon the requirement that a sequential relationship (for A to be a cause of B, A must occur before B) is a defining characteristic of causality, dismiss relativity theory as wrong, limit the applicability of the theory to the world of the subatomic particle (that is, claim that it does not apply to the macroscopic world of our experience), or, as will be done here, posit that *time has a single direction that we experience as "forward."* This supposition limits the applicability of our discussion to situations that occur at substantially below the speed of light, but this has the benefit of allowing the discussion to proceed. Therefore, our conclusions will not be relevant to events that occur at velocities approaching or exceeding the speed of light, although those types of events will briefly be mentioned in chapter 6. Another phenomenon that will not be dealt with here is *simultaneity*, a prediction of quantum mechanics that an event A distant from another event B can affect that other event B at the exact moment that A is occurring. Experimental evidence has already demonstrated that this occurs over a distance of meters, and the distances at which this phenomenon has been observed has been increasing in successive experiments over the last several years.

The prior few paragraphs, those positing the existence of unidirectional time and the exclusion of events near the speed of light or demonstrating simultaneity, are examples of the idea that the identification of the limits can actually encourage further discussion because it describes the boundaries within which ideas should be applied. This was the very concept used by Galileo when he rejected the Aristotelian model as being too complex. Doing so acknowledges that the conclusions that are drawn cannot be universal, but it has the potential of increasing the understanding of the concept under study.

Another outgrowth of quantum mechanics was Heisenberg's uncertainty principle, which states that it is not possible to know both an object's velocity and location at the same instant because the act of measuring one influences the other. This, too, has extraordinary implications for the study of causality since it indicates that an object's location in space cannot be determined with absolute accuracy. Said in a more general way:

The amount of information one can gather about any phenomenon is limited because measurement itself affects the process that is being described.

If one accepts that this principle of particle physics can be applied to the macroscopic world in which humans operate, then the implication is that causality can never be stated with 100 percent accuracy, because the act of determining whether A causes B influences the relationship between A and B.

Why accept the uncertainty principle but reject the reversibility of time introduced by relativity theory as applicable to our understanding of time and sequential, causal relationships? The answer is a practical one. We are looking for as universal a definition as possible and will make assumptions or limit the applicability of accurate concepts only when we must. A sequential relationship in time is a necessary component of the causal relationship. The claim that there will always be some uncertainty in claiming or proving a causal relationship, a point made by Hume, is irrefutable and does not prevent the discussion from proceeding. Some support for the application of the uncertainty principle to higher-order phenomena is demonstrated by Chaitin's extension of it to computer programs. The implication of the uncertainty principle most relevant here is that limits to knowledge, including causal knowledge, exist in any closed system.

A similar limit emerged in Kurt Gödel's (1906–1978) 1931 proof that it is not possible to construct a mathematical system in which all theorems can be derived. His incompleteness theorem states that every mathematical system will have statements that can be proven *only* by stepping outside that system. Because mathematics lies at the heart of the physical sciences—and if (and this is a big "if") one accepts the premise that all causal statements can be stated mathematically—then Gödel's incompleteness theorem can be restated as showing that it is not possible to describe or specify a physical system in which the cause of every event can be known.

In any closed system, the measurement (definition) of one variable will affect or influence the measurement of all other variables. This means that some uncertainty about the variables in any system is inevitable and unavoidable. As a result, there will be elements of every causal relationship that must be hypothesized or extrapolated because every such relationship cannot be proven.

The crucial implication for the present discussion is that both Gödel's incompleteness theorem and Heisenberg's uncertainty principle establish inherent limits that the scientific method cannot surmount. They also restate and parallel Hume's recognition of the limits of inductive reasoning, although on different bases. Since no system that utilizes mathematics or depends upon the principles of particle physics can be described or specified with total accuracy, it is impossible to prove that one has specified, much less accurately measured, *all* the causes in any closed system; knowledge or assumptions from an "outside" source will always be needed.

The power and influence of the scientific method during the twentieth century is demonstrated not only by advances in knowledge but also by the claims of practitioners of both psychoanalysis and Marxism that they are scientific disciplines. The Austrian philosopher of science Karl Popper (1902–1994) disagreed with this claim of both fields and sought to identify the central features of the scientific method in order to counter it. This brought him face to face with Hume's denial that cause could be absolutely determined; after all, if science differs from a field like psychoanalysis because science can discover true relationships in nature, then science should also be able to determine if the relationship between two events is causal.

Popper chose to emphasize one tenet of Francis Bacon's writings, proposing that the unique characteristic of science is the requirement that its theories or relationships can be *negated* or disproven. This idea is referred to as *falsifiability*. Like Bacon and Hume, Popper acknowledged that experiments that document a sequential relationship between two events support the claim of a causal relationship, but Popper believed his "discovery" of falsifiability answered Hume's conclusion that inductive reasoning (the unprovable leap) is always a part of causal reasoning. If Popper's claim were true, then an absolute distinction could be made between scientific and nonscientific knowledge and, by implication, between causality that is ascertained by scientific and nonscientific methods.

However, many activities and theories that the scientific community embraces as scientific do not meet Popper's criterion of falsifiability. For example, many past or one-time events are not amenable to experimental study, and this limits the ability to determine falsifiability. It may be possible to eliminate certain hypotheses by showing that they are untrue, but designing falsifiable tests is sometimes not possible. Evolution is an example of a theory that most would agree is scientific but for which the supporting evidence is primarily positive, not falsifiable. (Evolution as a causal concept is discussed in chapter 11.)

Popper's focus was the definition of scientific reasoning, not the identification of criteria that prove the presence of a causal relationship. Nonetheless, the elimination of alternative explanations through the construction of falsifiable hypotheses and analyses is a criterion that can be used to bolster claims of a causal relationship when it can be applied. The combination of multiple lines of positive evidence showing a relationship between two events and the elimination of other possible explanations through falsifiable comparisons offers strong support for a causal relationship. Counterfactual methods, which will be discussed at the end of chapter 8, are an application of this idea.

NECESSARY ASSUMPTIONS

This chapter has identified several thorny philosophical issues that are being dealt with by fiat or assumption. These assumptions underlie much of what follows, and readers who disagree with them will likely disagree with the ideas developed from them. Although making them explicit should help identify sources of disagreement, it will not settle them because they derive from differences in beliefs ("how we know"), not facts ("what we know"). Their resolution ultimately will rest on rhetorical (*verstehen*) rather than empirical (*verklaren*) methods.

Given their importance to the rest of the book, these assumptions or starting points will be repeated here:

1. The concept of causality is valid and describes a process by which one event brings about or increases the likelihood of the occurrence of another event.
2. Causes are discoverable, but absolute certainty about causal relationships is not possible.
3. Time is experienced as unidirectional, moving from the past to the future.
4. There are several models of causality. These complement rather than contradict or supersede one another. No single model of causality can claim that it is irrefutable, universal, or irreplaceable, including the one proposed here.
5. The choice of which model of causality to use is not random but depends upon the type of question being asked and the elements of the specific causal chain being considered.

SUMMARY AND REVIEW

This brief, selective historical review reveals that the concept of cause has changed dramatically over the four millennia of recorded history. Early writings and the belief systems of many cultures attribute to what we now call supernatural forces the ultimate cause of both the universe and of everyday events. In the West, Greek philosophers developed a rationalist approach that remains influential 2,500 years later. Included among these conceptualizations is the Aristotelian proposal that causality has multiple meanings. In the East, ideas of indeterminacy, feedback, and the circularity of time emerged at the same time. The flowering of the scientific method in the sixteenth and seventeenth centuries led to a focus on proximate cause and an emphasis on careful description and experiment. The demonstration by philosophers such as David Hume that there are limits to what can be stated by focusing primarily on proximate cause is reflected today in the existence of multiple models of causality and in the rejection of the existence or utility of causality by others. Each of these approaches has its advocates, many of whom believe that their approach is the only acceptable way of determining (or rejecting the existence of any) cause. What is often lacking is an acknowledgment that each approach has limitations.

Certainly, the conception of causality has changed over time. Sometimes this change has primarily been an emphasis of one idea rather than another, but the development of new technologies has been a spur to many of these changes. As this process has occurred repeatedly throughout recorded history, it is likely that conceptions of causality will continue to change. As a result, the meanings of causality and the methods by which knowledge is gained about it will continue to change.

2

THE THREE-FACET MODEL:
AN OVERVIEW

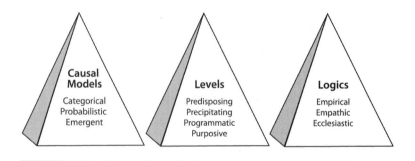

As chapter 1 illustrated, causality has been conceptualized in many different ways over the centuries. The current view of causality is that it has multiple meanings and that these vary by circumstance. The goal of this book is to present a model that brings the multiple ideas that characterize current views of causality together into an approach that recognizes the strengths and limitations of each, the complementary nature of their contributions to the overall topic, the duplication that is inherent in applying nonoverlapping but sometimes parallel approaches, the need for guidance regarding when to use a specific approach, and the fact that strict guidelines are not possible. This chapter will provide a brief overview of the model. Subsequent chapters will review in detail the facets and the concepts that underlie the model.

THE PROPOSED MODEL

The model proposed here can be described as a multiple-concept/ multiple-method approach that seeks to weave together three different aspects or facets of causal inquiry:

Facet 1: Three *models* of causes exist: the *categorical* (absolute or binary), the *probabilistic* (dimensional or continuous), and the *emergent* (nonlinear). The categorical model identifies causes that directly bring about an event, for example, an automobile accident leading to a whiplash injury, while dimensional causes influence the likelihood of an event. An example of the latter would be the wet road that made an accident more likely to happen that day. The categorical model involves "yes/no" reasoning, and the probabilistic model is reflected in the phrase "more/less likely." The nonlinear cause of the accident in this example would be the sudden occurrence of hydroplaning that occurred as the car accelerated on the wet road, reaching the speed at which the tires lost contact with the road and the driver lost control. It encompasses the entire set of circumstances (including, in this example, the coefficient of friction between the tires and the road, gravity, the amount of water on the road's surface, and the loss of momentum that occurred when the driver applied the brakes) that together bring about the sudden emergence of the outcome— loss of contact between the tires and the road and the resulting loss of control of the car, that is, hydroplaning.

Facet 2: Causality can be examined at four *levels* of analysis. First proposed by Aristotle but significantly modified here to fit modern conceptualizations, this facet assumes that the ascertainment of cause is best accomplished by analyzing factors at specific levels of analysis and that the choice of upon which level an analysis should take place is determined by the question being asked and by the characteristics of the issue whose cause is being sought. For some issues, multiple levels of analysis lead to the best understanding. The four levels of analysis are *predisposing* (factors that exist before the event occurs and increase the likelihood of its occurrence), *precipitating* (the necessary event whose close proximity to the onset demonstrates that without it the event would not have occurred), *programmatic* (the interactions among multiple elements that contribute more than any one of the constituent elements in bringing about the event), and *purposive* (the "why" an event occurred).

In the example of the accident, the excessive speed, the driver's tendency to become very emotional when under stress, the bald tires, and the heavy rain would all be predisposing causes. The excessive speed for those road conditions would be conceptualized as the precipitating cause, since without it the hydroplaning and therefore the accident would probably not have occurred. The programmatic causes would include the design of the road and the makeup of the road surface, which allowed water to accumulate and hydroplaning to become more likely; the characteristics of rubber and tread design that make bald tires vulnerable to losing contact with wet roads; and the characteristics of the driver and his situation that led to excessive speed. The driver's statement, "It's a miracle no one was seriously hurt; the accident was God's way of showing me that I have to take more responsibility for my actions," is an attribution of purposive cause.

Facet 3: Three distinct *logics* can be used to determine cause: the *empiric*, which requires that a question can be subjected to experimental study that can repeat, replicate, or in some manner validate the hypothesis or the data; the *empathic*, in which events are linked in a coherent, comprehensive, and convincing manner and in which the causal connections are understood to depend upon the personal understanding of the person or group making the connections; and the *ecclesiastic*, in which causal knowledge derives from a group-shared position of preexisting absolute knowledge.

In this example, the empirical evidence would be the demonstration that bald tires lose contact with wet surfaces at a much greater rate than tires with intact tread and that tread designs influence how much water is between the tire and the road at any given instant.

·

Other empirical evidence would be the results of experiments showing the relationship between the speed of a car and the coefficient of friction and at what point there is a sudden loss of contact. Empathic or narrative logic would weave together the facts that the driver of the car causing the accident was upset because he had just been given a poor performance rating by his boss and therefore was driving at a reckless speed in a severe downpour and that he could not afford to buy new tires because he was in personal bankruptcy. A claim that accidents are God's will and God's way of punishing one or of making a point would be ecclesiastic logic.

Table 2.1 presents my adaptation of the Aristotelian model and gives several examples of its application. It applies this schema to three events. In the collapse of the World Trade Towers on September 11, 2001, *predisposing* causes included characteristics of the material used in the construction of the building (steel melts in high heat), the failure of the material covering the steel to protect it from the heat of the fire, the failure of the intelligence community to

Table 2.1   The Four Levels of Cause: Three Examples

Levels of Cause	Chernobyl Nuclear Disaster	9/11 World Trade Center Building Collapse	Substance Use Disorder
Predisposition	Instrument panel design	Lax security FBI failure to follow up suspicious training Steel melts in fire	Genetic vulnerability Reinforcing property of drug Poverty
Provocation	Operator error	Hijacking Aviation fuel	Exposure Life stress
Programmatic	Nonredundant engineering	Building design	Biology of brain's pain and reward systems
Purposive	"Hubris"	Symbol of capitalism	Natural selection (pleasurable experience developed to increase the likelihood of survival and reproduction)

recognize that multiple hijackings were being planned even though a suspect who was learning to fly but not *to land* airplanes had been arrested in August 2001, airline security procedures that allowed box cutters and razor blades to be brought on planes, and a widespread belief that intrusive screening would discourage people from flying and undermine the aura of safety surrounding the airline industry. The *precipitating* causes of the towers' collapse were the hijacked airliners and the large amount of jet fuel they were carrying. The *programmatic* or system causes of the collapse included a building design that did not foresee the intense and prolonged heat that would be generated by a large amount of burning jet fuel and fact that the force of the collapsing upper floors could not be borne by a design in which the weight of the structure was concentrated in steel beams located in the center of the structure. Several *purposive* causes have been proposed, and their content depends upon the perspective of the person speculating. The perpetrators of the act claimed that the cause was Western imperialism against Arab nations and Islam and American support of Israel. Some commentators stated at the time of the incident that it was a punishment for widespread approval of abortion and other sinful acts in the United States. Still others claimed that the poverty and economic underdevelopment of many countries of the Middle East drove the planners and hijackers to carry out acts that would be both symbolic (explaining the specific buildings that were targeted) and destructive to a system that ignored the needs of the poor and both abused and exploited them.

Table 2.1 also applies facet 2 analysis to the Chernobyl nuclear reactor disaster, which occurred in the Soviet Union in 1986. The table lists the *predisposing* causes as a design in which valves allowed overheated water to enter into the reactor vessel and meters that implied that the overheated water needed to be removed from the reactor vessel. The *precipitating* causes were an incorrect reading of the instruments by the human operators when overheated water entered the reactor vessel and the operators' decision to stop water from entering the reactor. The *programmatic* causes were several design flaws that magnified the initial human error and did not provide the human operators with the information needed to understand what was happening. This was compounded by a lack of redundancy in the design of the system (elements that have been designed into all other designs of nuclear reactors) that could have either overridden the initial errors or not allowed the series of errors that occurred to be perpetuated and thereby evolve into a situation of "no return," that is, a situation that could not be reversed once the initial error was made. Proposed *purposive* causes include an ever-increasing need for electric power on a planet in which non-nuclear resources of energy are becoming scarcer and the claim by nuclear energy opponents that hubris has led humans to believe that nuclear energy can be harnessed with little risk of adverse outcomes.

The third example presented in table 2.1 is substance abuse. Among the *predisposing* factors are genetic vulnerability, which is a predisposing factor because many individuals with the genetic factor do not develop substance abuse disorder because of lack of exposure, thanks to unavailability, religious beliefs, social background, or an unhappy experience with a loved one who had a substance abuse problem. Other predisposing factors include a culture that simultaneously extols the virtues of drugs and casts them as villains, the easy availability of legal and illegal substances that have the potential to induce dependence and withdrawal symptoms, and social/demographic issues such as poverty, physical and sexual abuse, racism, and lack of knowledge. The specific pharmacologic properties that make a drug potentially abusable depend on its chemical structure and can be considered both predisposing and precipitating causes. For example, there may well be compounds existing now that are highly reinforcing and induce withdrawal symptoms when stopped after prolonged use and thereby meet the definition of addicting but are not drugs of abuse because they have not been "discovered." Their chemical structures are "predisposing" in that they contain the potential to cause abuse and dependence; the chemical structure is also precipitating because once exposure (a definite precipitating cause) has occurred, it is the structure that leads to or "causes" addiction and craving for further use and so induces abuse. The structure and function of the reward and pain relief systems that are "hard wired" in the brains of human beings and other species are also predisposing causes. These systems are "captured" by drugs of abuse and perpetuate the behaviors associated with taking the abusable substance. *Programmatic* causes of substance abuse include the societal factors that perpetuate their availability and encourage their use by making them illegal and therefore more exciting to some individuals. *Purposive* causes include the making of profit from growing or selling the drugs, the distribution of drugs by people who believe their use will destroy a culture that they disapprove of, the innate evil in human beings, and the search for pleasure by some individuals in spite of known risks.

Table 2.1 also demonstrates how the concept of causality depends on the knowledge base of the times. For Aristotle, the level of analysis labeled here as "predisposing" referred to inherent qualities. A statue preexists in the material and the sculptor is releasing it (an idea famously expressed by Michelangelo two thousand years after Aristotle), whereas the proposal here includes the sense of probability, a construct that did not exist in Aristotle's time.

This model is admittedly complex. It might help the reader to note that the four levels of analysis begin with the letter *P* and the three logics with the letter *E*.

3

THE ANSWER IS EITHER
"NO" OR "YES"

Causality as a Categorical Concept

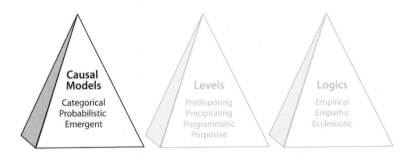

In a very short period of time, computers have become an indispensable part of everyday life, yet the idea upon which they are based is extraordinarily simple: knowledge can be coded and stored in a binary or *dichotomous* fashion such that each information point consists of a question with only two possible answers. In numeric representation the two possibilities are usually "zero" or "one," while in linguistic representation the two choices are usually stated as "no" or "yes." The computer represents these two states in an electronic circuit in which a switch is either open or closed. When the circuit is closed and the current flows, the value of one or yes is usually assigned, while zero or no is indicated when the switch is open and no current flows.

The widespread availability of computers was made possible by the invention of the transistor and the subsequent development of techniques for inexpensively manufacturing microscopic chips. However, the basic concept of binary logic has been used by humans for thousands of years. Because the two possible states of the binary approach are mutually exclusive (each information point can only be one of the two possibilities) and absolute (it must be one or the other), this approach is referred to as *categorical*, binary, digital, or dichotomous. These words will be used interchangeably throughout this book when the form of logic that follows this pattern is being discussed, even though there are subtle differences among them.

Categorical logic is part of our everyday discourse and is the usual model of reasoning used to conceptualize causal relationships, at least in the West. When we want to know why an airplane crash occurred, why a teenager shot several schoolmates, why the Soviet Union fell apart in 1991, or why computers became widely available during the last two decades of the twentieth century, the question is usually framed as seeking a single or primary reason. When we are told that a worn rudder screw caused the plane crash, that bullying led the teenager to seek revenge, that massive military spending by the United States caused the disintegration of the Soviet economy because it could not keep up, or that the development of the compact, inexpensive modern computer was made possible by the discovery that chips with thousands, millions, or even billions of very small circuits could be inexpensively manufactured on thin wafers of silicon, we are using categorical logic. Each of these answers is clear cut, readily understandable, and unequivocal. These general features of the dichotomous, binary model are among its strengths, and they make this approach to causal logic appealing and widely used.

However, these examples also expose the limitations of this approach. Each of these events probably had several or many causes and influences, and the relationships among them are likely complex and sometimes difficult to enumerate. Even if we cite less complex events, such as the joining of two boards that follows a nail being hit with a hammer, improving grades after studying harder, and the collision of two automobiles at an intersection occurring after one car does not stop for a red light, we must choose the level at which we want to analyze the event—interacting molecular forces, interacting objects (nail and wood, automobiles), or the purpose of the person initiating the event (building a shelf, not wanting to be late for an important meeting)—before coming to a final conclusion.

In fact, because this question about which level of analysis to use can be raised about any event in which causality is being examined, the categorical approach to causality is sometimes dismissed as simplistic and out of place. Yet categorical logic is hard to get away from. Not only do we use it to discuss causality in our everyday lives, but we depend on it make decisions, to engineer the mundane and the magnificent, to act in an emergency, and to plan for the future. These successes attest to the utility and importance of this approach and make it a natural starting point for an analysis of the concept of cause. Perhaps the widespread use of binary logic derives, as Kant suggested, from the fact that our brains are organized to see things categorically, but perhaps our behavior reflects a fundamental way in which nature works. In this chapter, we will put aside the causal question of why it is so widely used and focus on the strengths and limitations of this approach. Doing so will clarify when the categorical, dichotomous approach is applicable and when other approaches are better used. Subsequent chapters will have the same goal in relationship to other approaches to causality.

One reason to begin with the categorical approach is that its simplicity can help enumerate the properties needed to demonstrate a causal relationship. The most straightforward example of a categorical causal relationship is when an event B occurs only after another event A occurs. This is an example of A's being *necessary* for the occurrence of "B." If B occurs every time A occurs, the relationship is referred to as A's being *sufficient* for B to occur. This approach to causality, that is, enumerating the necessary and sufficient causes, was first used by Galileo, and it is the essence of Koch's postulates and of our understanding of physical phenomena such as gravity. B occurs only if A occurs first. Huntington disease occurs only when a specific genetic abnormality is present, and objects "fall" because there is an exchange of particles between them when they are placed or brought into a certain spatial relationship. To generalize, one characteristic of causal relationships is that the events occur *sequentially*, that is, the reputed effect B follows *after* the reputed cause A.

A second and related characteristic of causal relationships is that A and B should be related *temporally*, that is, have some proximity in time. This proximity could range from nanoseconds to millennia, depending on the nature of the event in question, but it must be compatible with the context of the event that is being explained.

Third, if the event being explained has occurred more than once, the temporal relationship should also show *regularity*, that is, B should often occur after A. If we limit ourselves to the necessary and sufficient relationship, then B never occurs without A, but if A is merely sufficient for B, then the likelihood of a causal relationship is increased if the sequence is observed multiple times, hence regularly. As a corollary, the demonstration of a causal relationship becomes much more challenging if an event has occurred only once.

The criteria of sequentiality, temporality, and regularity apply even when the relationship between a cause and effect is not necessary and sufficient, but these criteria and hence a causal relationship are most easily demonstrated when a situation can be manipulated experimentally or when events occur repeatedly. For example, Koch was trying to determine whether a specific infectious agent causes a specific disease. He and others developed methods by which a regular relationship could be demonstrated in an experimental (artificial) situation that could be repeated in such a fashion to demonstrate a replicable and unique or singular relationship between an organism and a disease. However, because many situations we are interested in have already occurred, are not reproducible in a laboratory situation, occur only once, or recur in very different circumstances, these criteria cannot be satisfied in many situations.

The major limitation of the "necessary and sufficient" requirement, though, is the requirement of exclusivity or specificity—that is, that B occurs *only* when A is present. Indeed, this is impossible to prove since one could never prove that all examples of B have been examined. Thus, the concept of "necessary and sufficient" may be useful in the abstract, but such a relationship can never be proven with absolute surety.

This limitation returns us to the concern that Hume raised more than 250 years ago: even when a relationship between A and B is "obvious," inductive logic is being used. We can neither guarantee that every occurrence of B is preceded by A or that B inevitably follows A because we can never be sure we have identified every instance of either A or B. Thus, as Hume noted, a causal relationship can be inferred but not conclusively proven. The likelihood of a causal relationship may be so high that we can be very confident of a causal relationship, but we must accept Hume's conclusion that "extremely likely" doesn't equate to "absolutely proven."

Plausibility as a feature of cause was suggested by the British epidemiologist Austin Bradford-Hill, whose contributions will be discussed in more depth in chapter 8. Plausibility implies that other evidence is available to support a direct causal relationship. However, it seems more subjective than the other criteria because it is less easily operationalized. Plausibility does share with the criteria of sequentiality, temporality, and regularity the requirement that people (often experts) agree that they are present. This raises a broad question that will arise at several points in this book: is causality a social construct, or does it have an essence that exists outside of the human mind? The noted philosopher of science Thomas Kuhn (1922–1996), for example, suggested that consensus in science often comes about only after those who hold power are replaced by younger individuals. This implies that there is a scientific "truth" but that at any one time what is accepted as scientific truth is affected in part by social forces. A more radical representation of this view, postmodernism, claims that concepts such as "truth" and "cause" should be abandoned because they are merely social constructs. This is ultimately a question of belief, not proof; hence, the statement in the introduction that we begin with the *supposition* that causality exists rests upon a belief that it is an actual aspect of the natural world as well as upon its utility as a concept. The criterion of plausibility illustrates that the determination of causality has a subjective element and that some criteria are more readily operationalized and therefore less likely to generate disagreement than others. Plausibility was likely one of the reasons for the quick acceptance by many scientists of Darwin and Wallace's theory of evolution. However, two other examples of scientific theories that we will review later, the tectonic plate theory of geology and Mendel's theory of trait inheritance, were dismissed for many years in part because their plausibility was not obvious to many, suggesting that a lack of perceived plausibility can delay acceptance. This suggests that plausibility can play a permissive role in establishing the presence of a causal relationship but should have less weight as a primary criterion. Furthermore, plausibility plays a role in identifying possible causal relationships and in suggesting testable hypotheses. Perhaps its most important role is to spur the search for other data or experiments that increase plausibility and thereby lead to the development of further evidence supporting a hypothesized causal relationship.

Simplicity is another criterion sometimes used to support a theory's validity. This concept is alternatively referred to as "Occam's Razor," named after a fourteenth-century English monk who stated that one need not think of multiple things when one will suffice. This is often restated as favoring the simpler, more straightforward choice when more than one plausible explanation exists. Adjectives such as "elegant" and "economical" are often used to describe results or theories that meet this criterion. While the subjectivity of simplicity is obvious, its use as a criterion rests on the belief or observation that "nature" has often achieved an end through a very efficient means. This would seem to lack potent support as a characteristic of causal relationships but offers the useful idea that simpler explanations be considered before more complex ones.

The categorical approach has several major strengths. First, it has "face" validity, that is, it seems to confirm everyday experience. When we hit a nail with a hammer, the nail appears to enter the piece of wood. To extrapolate that hitting the nail with the hammer caused the nail to enter the board seems mundane. Second, a single yes/no answer lends itself to manipulations that can lead to observable results; if we believe A causes B and wish to study this hypothesis, we can induce A and observe if B occurs. Or, if we wish to lessen or stop B, then we can determine if changing A often brings about the desired result. Third, experimental data suggests that the concept of causality is innate and that the yes/no answer has neural concomitants in the brain. This can be sited as evidence of external validity of the construct of binary causality, although it also confirms Kant's idea that we are constructed to see the world in this way. Fourth, the categorical approach is straightforward and easily understandable. Occam's razor would favor its acceptance.

IS THE BINARY MODEL TOO SIMPLE?

What about the protestation that life is not as simple as the binary model implies? Certainly, many events have multiple causes or require that several preceding factors be in place before they can occur. While this may seem obvious, the complexity of establishing causal relationships between *two* events is challenging enough; the determination of a complex web of causes and effects almost always requires a concerted effort over an extended period of time. It shouldn't be surprising, therefore, that many advances in science come about by testing one idea at a time. Only later are the results of multiple experiments and multiple hypotheses put together to establish that several or many causal contributors are necessary. In fact, many well-designed experiments "control for" all factors that the researcher does not want to examine at that moment or is uninterested in. This is just what Galileo hoped for four centuries ago when he rejected the complex Aristotelian schema for a simple causal model that focused on proximate cause.

In the simplest example, some causal sequences require the linkage of a number of binary events. If event A is shown to cause event E, this might depend upon A causing B, B causing C, C causing D, and D causing E. Each relationship can be analyzed as a binary occurrence. In this instance, the simplicity of the binary approach is a strength. It simplifies a complex situation and leads to the determination that simple models can be amalgamated over time into more complex models while still allowing them to be tested with categorical methods.

However, many events occur *only* if two or more events have occurred previously, if two or more events occur coincidently, or because the occurrence of one event increases the likelihood that another will occur. Many other possible scenarios exist. For example, the coincident occurrence of three events might be necessary. Judea Pearl, in his book *Causality*, presents a notational method for identifying many of the possibilities and for teasing apart the complexities of causal relationships.

Clearly, the categorical approach is limited in a number of ways. As we've seen, many events appear to have more than one cause, so a single yes/no answer is often not correct. Further, since the "necessary and sufficient" requirement cannot really be demonstrated, it is not possible to claim exclusivity or specificity of relationship as implied in binary causality. Third, if cause can and sometimes should be analyzed at several levels, as Aristotle suggested, one can reasonably ask if each level of analysis requires a categorical analysis. Even if the answer is yes (I will argue later that the answer is "sometimes," since analyses at multiple levels often invoke noncategorical logics, techniques, and data), the analysis requires more than the single yes/no answer required by the standard binary approach. Fourth, the yes/no analysis of the dichotomous approach does not fit every situation. Many causal relationships depend on the amount or degree or timing of one or several components or on the co-occurrence of several events. These limitations suggest that the categorical approach cannot be the only model. The question, therefore, is not *whether* the categorical model is ever or always accurate or useful but rather *which* events of interest can usefully and most accurately be understood using this model. What are the alternatives? The next two chapters review two alternative models, the probabilistic and the nonlinear.

4

PROBABILITIES, ODDS, AND RISKS

Predispositions and Provocations as Causes

It is better to have an approximate answer to the right question than an absolute answer to the wrong question.

—John Tukey

Events may happen strictly by chance and still be obligatory.

—Christian De Duve

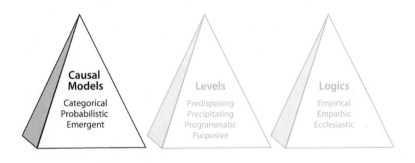

Chapter 3 presented the concept of cause as categorical. In that binary or digital model, something either *is* or *is not* a cause, and, if it is, it acts directly to bring about an event. This chapter presents an alternative model, the probabilistic approach, in which causes are conceptualized as *events that affect the likelihood that another event will occur*. In this model, causes act as influences, risk factors, predispositions, modifiers, and buffers.

The complexity of the probabilistic concept of cause begins with the definition of the word "probability." Its primary meaning relates to predictability or prognostication, that is, the *likelihood* of specific future outcomes or effects. However, this definition has within it the implication, and some would claim a second meaning, that there is *uncertainty* as to the outcome. This implication or second meaning acknowledges from the start that there is an inherent limitation in the ability to predict likelihood and therefore a limitation in the certainty with which a causal relationship can be identified. Werner Heisenberg, the physicist who first described the uncertainty principle, captured these two elements quite succinctly when he stated, "The probability function contains the objective element of tendency and the subjective element of incomplete knowledge."

What are the characteristics of probabilistic reasoning that are relevant to causality? First, probabilistic logic is *dimensional*. That is, causal probabilities can have any value between zero and one, that is, between 0 percent and 100 percent. However, they can never be either zero or one because that represents categorical cause. This raises the challenge of interpreting the meaning of a specific value. For example, what does a 0.2 probability or 20 percent chance actually mean?

A second feature that follows from the dimensional nature of the model of causality is that the likelihood of a causal relationship changes in a *regular, graded* fashion. That is, 10 percent is twice as likely as 5 percent. This is important because it allows one to distinguish between minor or insignificant differences in likelihood (say between 23 and 26 percent) and large differences (say between 25 and 75 percent). This regularity is a crucial feature of the probabilistic approach because it allows for mathematical manipulation, the ranking of the probabilities of different events, and for an intuitive appreciation of differences ("twice as likely" or "ten times as likely").

However, what seems like a minor difference can have a very large effect if the number of events or organisms involved is large or the time scale is long. For example, if the blood pressure of the entire population could be lowered by a mere four points on average, the number of strokes that would be prevented would be greater than the number prevented by treating everyone with "high" blood pressure; with compound interest ten dollars becomes $100.63 in thirty years at 8 percent interest but only $76.12 at 7 percent interest.

Not all causal models describing relationships between events as mathematical likelihoods are linear or regular, however. Chapter 5 will discuss the nonlinear model of causality, in which probabilities do not change in this regular, graded fashion.

A third characteristic of probabilistic causality is the aforementioned *uncertainty* that is inherent in all probabilistic statements. Stated alternatively, probabilities have ranges rather than absolute, invariant values. The range of uncertainty can be narrowed by collecting more information, but it is never zero (no uncertainty).

Thus, the probabilistic model of causality provides a description of the likelihood that two events are associated, not an absolute statement that there is or is not a causal relationship. It indicates possibility or probability, in contrast to the certainty that the categorical model represents.

To many people, the differences between the "yes/no" approach of the binary model and the "maybe" of the probabilistic model imply that probabilistic causality is both less informative and more complex than the categorical model. For some, phrases such as "might be related causally" suggest indecisiveness and lack of precision. Ironically, the converse is true: probabilistic descriptions maximize the amount of information that can be conveyed about a potential relationship between events, even though they acknowledge the inherent uncertainty. Perhaps this incongruity between the amount and specificity of information provided by probabilistic methods is the reason that even today many people do not think about cause in a probabilistic fashion or, when they do, think about it in unrealistic ways.

Whatever the reason, probabilistic reasoning and its application to causality are difficult for many to grasp and apply. A brief diversion into the historical development of the concept can help explain why.

The basic idea underlying probabilistic thinking seems to have emerged 1,800 to two thousand years ago. In his book *The Science of Conjecture*, James Franklin identifies discussions in the Sanskrit Laws of Manu, the Roman Digest, and the Jewish Talmud, all of which date from that period of time, as the first expressions of probabilistic reasoning. Each discusses how to apportion responsibility in those legal cases in which the evidence of wrongdoing is not absolute. Franklin argues that the lack of any mention of probabilistic reasoning in writings before then indicates that the ancient Greeks were unaware of the concept of probability and adds that the absence of the concept of incomplete legal evidence in many cultures and the lack of a word for probability in Middle English suggest that probabilistic reasoning is absent in some cultures and therefore not a universal form of human reasoning.

This conclusion is indirectly challenged by recent research that has found that probabilistic causal reasoning can be carried out by nonprimate animal species. For example, Aaron Blaisdell and colleagues did a series of experiments in rats showing that animals who had never done a task could predict the likelihood of an outcome after observing other rats performing the task. Blaisdell designed the experiment to rule out mimicry or learning as the explanations and demonstrated that the reasoning was probabilistic in nature; that is, when they experimentally increased the likelihood of an outcome, they increased the likelihood that the rats would act in a certain way. These experiments demonstrate that probabilistic reasoning is inherent to rats and suggest that the rat brain is wired to make probabilistic calculations. The researchers extrapolate to claim that probabilistic reasoning must also be inherent in humans.

If human brains are "built" to carry out probabilistic reasoning, then the lack of conceptual or linguistic representation of the concept in some cultures and epochs is puzzling. Indeed, scholars agree that the first explicit discussion of probabilistic logic and its application to everyday problems did not occur until 1652 in the *Port-Royal Logic*, published by residents of a Parisian monastery (who were likely significantly influenced by the mathematician Blaise Pascal). It introduced two ideas essential to probabilistic reasoning: the concept of the ratio as a means of expressing the likelihood of specific outcomes in games of chance and the idea that knowing the likelihood of an event's outcomes increases the ability to predict a correct answer compared to guessing, though not with absolute certainty or accuracy.

The *Port-Royal Logic* was quickly followed by the Dutch polymath Christian Huygens's textbook on probability and the British mathematician John Graunt's development of methods for calculating the likelihood of mortality using samples of data collected in church and town ledgers. They thus both expanded the application of probabilistic logic and made explicit the dramatic differences between categorical and probabilistic reasoning.

This brief history presents a conundrum. If the concept of probabilistic cause is inherent, as suggested by the Blaisdell experiments, how can the ideas be alien to so many? How could probabilistic cause not be mentioned before the years 1 to 200 CE and not be explicitly discussed as a construct before the *Port-Royal Logic*? One plausible explanation is that both the mathematical tools required to express and quantify probabilistic logic and the concept of uncertainty had to be developed before the construct of probabilistic thinking could be made explicit. Among the necessary mathematical ideas needed to specify likelihood are the concepts of the zero and infinity. Zero was not mentioned as a placeholder in a mathematical sense until the third century CE in India, even though the Babylonians and the Greek philosopher Democritus had written about the quantities of none and infinity. Another needed advance was a numbering system that is continuous and infinitely graded. Roman numerals can express small or large numbers but cannot provide a representation of infinitely large or infinitely small because there is a maximum number (M = 1,000; many Ms are still not infinite) and a minimum number (I = 1).

Not until the tenth and eleventh centuries did Arabic-speaking Persian philosopher-scientists use zero in calculations and definitively apply the continuous numbering system currently in use. This numbering system allowed for direct computation beyond addition and subtraction by providing symbols that follow a regular, predictable pattern, for example, 1, 10, 100, 1,000, a clear advance over representational systems such as Roman numerals (I, V, X, etc.). Furthermore, the establishment of a numbering system that is *continuous* from zero to infinity (and later expanded to range from negative infinity to positive infinity) and *infinitely graded* (1, 1.1, 1.11, 1.111, . . .) demonstrated the utility and application of numbers that do not express exact quantities. This moved mathematics beyond the concept of complete accuracy that underlies categorical causality and allowed for the development of the methods and tools to express, in a practical manner, the uncertainty inherent in the probabilistic model.

This system of mathematical representation, then, provides support for both the abstract concept of probability and probabilistic causality and the means by which it can be represented in a concrete mathematical form. I will argue below that the development of more complex mathematical approaches such as the calculus and the concept of the normal distribution provided further conceptual and practical structure to the idea that uncertainty is not only quantifiable but can be mathematically manipulated in ways that allow the strength or likelihood of causal associations to be described.

Nonetheless, even today probabilistic reasoning and its application to causality are difficult for many to grasp and apply, a challenge apparent to the writers of the *Port-Royal Logic*: "Fear of harm ought to be proportional not merely to the gravity of the harm, but also the probability of the event, and since there is scarcely any kind of death more rare than death by thunderstorm, there is hardly any which ought to occasion less fear." Yet this fear is common and sometimes incapacitating. Thus, these authors knew that the widespread fear of being struck by lightning is far more common than one would expect based on the probability of death by lightning, a fact that seems to be just as true today, in spite of the fact that we know there are even steps one can take, such as not standing under trees or lying down if caught in an open field during a thunderstorm. That is, factors other than its statistical likelihood must causally mediate this fear because even the knowledge that there are mitigating actions hasn't mitigated the worry.

Another mid-seventeenth century advance that influenced the development of probabilistic causality was the invention of the calculus simultaneously by Newton and Leibniz. Calculus utilizes the concept of successive approximations to describe shapes and curves such as arcs and parabolas and to calculate nonlinear paths such as the flight of a cannonball. The method of calculation by successive approximations is called the limit theorem, which shows that a parabolic line approaches but never touches a limit. By summating smaller and smaller areas within the figure, calculus provides a measure of the area within the parabola that gets closer and closer to the "actual" area—even though a small area is always unmeasured.

The influence of calculus on the construct of causality derived from its ability to quantify situations and make predictions (calculations) when an infinite number of mathematical operations would be needed to get the "correct" answer, for example, the amount of physical stress operating at various points in a tall building. In doing so it freed engineers, designers, and scientists from the need for total accuracy in situations in which that is impossible to achieve and, as such, provided a powerful alternative to the absoluteness of the categorical model. That is, it demonstrated that one can get so close to a correct answer that what is obtained cannot be distinguished from the actual and thereby showed that a mathematical result need not be absolutely precise for it to be accurate and useful. Calculus thus provided a method that embraces both meanings of the concept of probability identified by Heisenberg: a mathematical representation of the likelihood of an outcome in a specific situation *and* the uncertainty inherent in that calculation. By embracing both meanings of probabilistic reasoning and allowing for quantification, it further united the theoretical and applied concepts needed for a science of probabilistic causality.

It is clear that the concept of ratios and odds, the calculus, and the quantification of probability developed around the same time. A number of potential explanations for this coincident development seem plausible: the concepts are related, the idea of mathematical prediction underlay all three, the development of a merchant class led to profits that increased leisure time and therefore an interest in topics such as accurate prediction in games of chance (gambling), and the emergence of universities encouraged individuals to consider a variety of measurement issues and to communicate with other experts. The almost simultaneous emergence of these three disparate but related ideas suggests some connection. However, it could well be the result of several or all of these explanations.

Another "tool" that significantly influenced both the development of methods to quantify probability and to influence the conceptual underpinnings of the current concept of causal probabilistic reasoning was Abraham de Moivre's (1667–1754) discovery that the distribution of seemingly random events follows a "bell curve" or normal distribution. This finding was further developed by Carl Friedrich Gauss (1777–1855), who showed that variability from the mean decreases as the number of measurements or samples increases, a finding that led to the conclusion that accuracy increases as the number of observations increases. This approach was then applied by the nineteenth-century polymath Francis Galton (1822–1911) to many measurable quantities, including the height and weight of humans. By showing that the results distributed themselves in a "Gaussian" distribution in large numbers of people, he helped foster the development of methods for quantifying such predictions. The ubiquity of the normal distribution further cemented the reality of probabilistic causality because it implied that there must be forces inherent in nature that lead to this probabilistic pattern of distribution. Galton also recognized that a major limitation of the probabilistic approach was its limited ability to make predictions about individuals.

The practical applications of outcome prediction and quantification also spurred the development of probabilistic statistical tools for improving the quantification and analysis of large amounts of data. The business writer Richard Bernstein goes so far as to argue in his book *Against the Gods: The Remarkable Story of Risk* that the concepts of risk and prediction and the development of methods for quantifying them were necessary precursors of capitalism and large industry. That is, they were not sufficient by themselves to have caused the emergence of capitalism but greatly increased the likelihood that it would emerge because, according to Bernstein, they allowed for a calculation of the ability to make a return on investment with the quantification of the risk of a venture, a central underpinning of capitalism. Bernstein is thus suggesting that one of the causal underpinnings of this powerful economic system was the conceptual emergence of probabilistic logic. Another aspect of the probabilistic approach that has entered human experience to such an extent that its relatively recent origins seem surprising is the calculation of the "average" (or "mean") that was developed by de Moivre in his work on the normal distribution.

Many examples of probabilistic causal logic have become part of the fabric of modern life. For example, speed limits lower the likelihood of an accident but do not eliminate them; good poker and bridge players know which cards have already been played and make subsequent moves accordingly, knowing that this increases but does not insure the likelihood of their winning; and successful athletes and coaches rely on their knowledge that certain actions by the opposition are more likely and try to anticipate them to increase their chances of winning. However, these actions are often perceived to be categorical in nature. For example, going over the speed limit is considered wrong because it is dangerous rather than recognizing that speed limits both lessen accident rates and decrease the risk of a fatal outcome if an accident does occur but do not eliminate either.

This brings us back to one of the challenges of causal probabilistic logic, illustrated in the following example. The likelihood that a person who smokes more than one pack of cigarettes per day for fifty years will develop lung cancer is about seventeen times greater than someone who does not smoke at all. This ratio has a "confidence interval," that is, a calculable range within which one can be "80 percent" or "95 percent" sure that this statement is accurate. However, even at the 95 percent confidence level, there is a one-in-twenty chance that the results will fall outside the range; at the 80 percent confidence interval, there is a one-in-five chance. This tells us is that the likelihood that cigarette smoking caused lung cancer in an individual smoker is much greater than zero—that is, much greater than chance—but, because there are other causes and because not everyone who smokes will develop cancer, one can only indicate a range of likelihood that it was causal rather than an absolute "yes" or "no." Nonetheless, this likelihood is so high that we can be quite confident that someone who smokes regularly is more likely to develop lung cancer than someone who did not, but we cannot be absolutely confident—and we cannot predict with any accuracy whether a specific individual will or will not develop lung cancer.

Does this prove that cigarette smoking is a cause of lung cancer? No and yes. On the "no" side, these data show only that there is a very strong association between cigarette smoking and cancer. As Hume pointed out, this does not prove a causal relationship. However, given the strength of the relationship (seventeen is a large relative risk), the probability that lung cancer can be caused by smoking is extremely high. In general, then, the probabilistic model is unable to establish absolute causal relationships. However, this does not mean that prediction is impossible or that it is fruitless to try.

Likewise, the risk that a person who smokes two packs of cigarettes daily for thirty years will develop cancer is clearly greater than the risk that a person who never smoked will. This uncertainty was seized upon by cigarette manufacturers and their supporters for years in their statements that there was no proof that cigarette smoking caused cancer, but the strength of the association ultimately convinced public health officials, juries, legislatures, and much of the public that the answer is "yes."

CAUSALITY AND PROBABILITY

The uncertainty inherent in probabilistic reasoning is seen by some as a reason to reject it, but the great contribution of the probabilistic approach is that it provides a method by which uncertainty can be quantified. It is worth noting that probabilistic reasoning is used in many day-to-day decisions that are not directly linked to causality, for example, the decisions to take along an umbrella, to choose between putting money into a certificate of deposit or a stock, accept a new job offer, or purchase a specific car. Furthermore, some decisions are made even when it is clear that the likelihood of an outcome is low. Examples include choosing to undergo a medical procedure even when it is unlikely to result in the desired outcome, investing money in a very risky venture, or continuing to smoke or engage in dangerous behavior. The strength of the probabilistic method is that it gives the decider a quantification of the risk involved. When applied to causal reasoning, the probabilistic method conveys information on the likelihood of a causal relationship while indicating there is no absolute answer.

LIMITATIONS OF THE PROBABILISTIC MODEL

As we have noted, the probabilistic model differs radically from the "yes/no" approach of the binary model. Probabilism indicates "maybe," whereas the categorical model represents certainty; probabilism provides a description of the likelihood that two events are associated, not an absolute statement about their relationship. Thus, probabilism seems both more complex and less informative to many people.

Several limitations of the dimensional, probabilistic model have already been mentioned. First and foremost, as just noted, it cannot identify causes with absolute certainty. This is a source of frustration to many and can be used to distract people from acknowledging that likely causal relationships exist. Second, for many, the probabilistic model is less intuitive than the categorical model; this undermines its ability to provide meaningful information and increases the likelihood that the information provided will be misinterpreted.

A third limitation is that the establishment of probabilistic information usually requires that a phenomenon occurs more than once, or that it affect groups of people or objects, or at least that there is knowledge about a similar circumstance. Because probabilistic methods are limited in their ability to provide likelihoods about single events or individuals, they are rarely useful in assigning causal information about unique occurrences such as historical events, specific evolutionary paths, the ecology, individual lives, or individual molecules. Very often, the unique aspects of any individual object or person so outweigh any similarities shared with others that causal prediction is close to or at a chance level.

This dependence on group data illustrates another limitation of the probabilistic model. Its ability to predict or describe the likelihood of a causal relationship depends on the equivalence of the events being compared. Since the uncertainty principle demonstrates that it is not possible to determine if any two events are exactly the same, some of the error in the probabilistic model derives from the inherent inability to determine the equivalence of the events being compared. Another source of error is that measurements can never be exact, no matter how careful the collector or measurer is. Thus, some of the "fuzziness" of probabilistic information derives from the fact that there are multiple reasons that the information on which an estimate is made has error and that this error cannot be fully eliminated. It was this realization that led Ronald Fisher to develop the science behind sampling, as discussed in chapter 8.

However, the probabilistic approach's inability to make absolute statements about causality or future actions should not be used as the basis for a claim that causality either does not exist or cannot be determined. A lack of absolute certainty underpins every approach to knowing except, as will be argued in chapter 10, the religious or ecclesiastic. The nihilism and outright rejection of causality that have resulted from this inherent uncertainty are themselves misapplications of categorical reasoning and a failure to accept the fact that there are limitations of human knowledge. The probabilistic model developed from an appreciation of these limitations and the realization that some information could be gained but that it cannot be absolute. While this is an inherent limitation of nature (or at least of how humans are constructed to apprehend nature), a recognizing of this limitation has actually improved the ability to use the information that can be gleaned. Identifying the limitations of each approach actually maximizes the ability to derive causal information because it increases the likelihood that a given method or approach will be applied in appropriate situations and lessens the likelihood of inappropriate or misapplication.

Finally, for many people, probabilistic reasoning does not seem natural or is not within their everyday experience. The higher prevalence of fear of flying compared to fear of riding in a car is an example. The likelihood that an individual will die in an automobile accident is much greater than in an airplane crash. Whether this greater fear of flying is the result of the dramatic nature of airplane crashes, the fact that many more people are killed at once in commercial airplane accidents, the fact that automobile drivers believe that they are more in control of the danger (true or not), or some other factor or factors is not clear. Yet many humans behave as if the probability of death by flying is much greater than death by automobile, suggesting that probabilistic comparisons don't always come "naturally" and that many individuals act in ways that defy the logic that derives *only* from probabilistic reasoning.

ARE PROBABILISTIC AND CATEGORICAL REASONING DISTINCT?

Thus far, categorical and probabilistic causality have been presented as distinct models of reasoning. Would they better be considered a single approach, with one being a subset of the other? Could not the categorical model be considered a special case of the probabilistic approach in which the probability of a causal relationship is either 100 percent ("yes") or 0 percent ("no")? Or, alternatively, why not consider the probabilistic approach a subset of the categorical?

Support for collapsing them comes from the dramatic success of the computer and the digital camera, both of which illustrate that complex, graded phenomena can ultimately be coded digitally. Perhaps nature is constructed digitally (categorically), while humans are constructed to perceive it continuously. Or, conversely, perhaps nature functions continuously, but humans have constructed categorical concepts to simplify it. Several reasons suggest to me that it is worth keeping the distinctions between the two models, at least at the conceptual level.

The inherent simplicity of the categorical model and the empowerment that this provides is a great strength. In applied fields such as engineering, the ability to talk and work in distinct categories enables action. This seems to be true in everyday life as well. For example, meteorologists distinguish between tropical storms and hurricanes. The fact that one has winds less than seventy-five miles per hour and the other has winds more than seventy-five miles per hour seems to make a difference in the attention paid to it by the public and by the preparations that are made. Simplicity was one of Galileo's justifications for abandoning the Aristotelian approach, and many advances in scientific understanding have resulted from focusing on single causal elements.

Conversely, the concept of multiple, interacting causes is easier to understand and to describe mathematically in the probabilistic model. Furthermore, as noted at the end of the prior section, the probabilistic model has led to an appreciation that an inability to identify one or several causes with absolute certainty does not mean that one is without any information about causal relationships.

What would be lost by subsuming or abandoning either categorical or probabilistic reasoning is the contrast between absolute and relative certainty. This distinction has both philosophical and practical implications; it avoids the claim that one approach is always more accurate or better than the other, and it is necessary for the claim made in this book that the choice of a specific model is determined or at least strongly influenced by the circumstances or events being considered. The categorical model favors single causes. Several categorical causes can coexist, but the concept of multiple, interacting causes is easier to understand and describe mathematically with the probabilistic model.

Abandoning the distinction between the two models would ignore the limitations of each. Furthermore, attempting to do so would be unsuccessful, in my opinion, since many people would continue to think and act in one mode or the other. Keeping both models encourages a recognition of the strengths and limitations of each and acknowledges the reality of human behavior. Said differently, we seem to think about causality in these different ways, and as long as people tend to do so, it is both useful and important to recognize that these different ways of approaching causality have very different implications. On the other hand, keeping them distinct (a categorical act itself) does not ignore the common features they share. Keeping them distinct also addresses Hume's insight that absolute certainty about causal relationships is not possible. By providing an alternative approach, the probabilistic approach allows for action based on causal reasoning even without 100 percent certainty. I suggest that Hume's objection reflects a limitation of the categorical approach but that action sometimes requires taking that approach nonetheless.

Another practical distinction is that the two models require different statistical approaches. The categorical model is tested using methods, called *nonparametric*, that can compare only several groups. In contrast, the dimensional model relies upon *parametric* approaches that compare features of distributed information such as means and standard deviations. Statistical assessments in the probabilistic approach are based on comparisons of ratios that are indicators of the likelihood that the distribution differs from chance.

Finally, the value of keeping the distinction between the categorical and probabilistic is supported by the discovery that different aspects of physical and biological systems are best described in either categorical or analog terms. Each of the following examples illustrates an aspect of a natural system that operates in a binary fashion but whose actions, when summed or examined from a more macroscopic approach, appears graded.

- Many subatomic particles act as if they exist in only one of two energy states (a categorical model) but follow probabilistic, dimensional modeling when examined in aggregate.

- In the brain, many ion channels, receptors, cells, and neuronal circuits are either active or quiescent (categorical) but function together with many other like features to operate in a graded, dimensional fashion (probabilistic).

- Individual neurons have been identified in the frontal lobes of monkeys that fire when monkeys move their eyes to a target. Other cells in the visual cortex fire when a line that has a specific degree of slant is placed in the visual field. These are categorical actions; each neuron either fires or doesn't. However, eyes appear to move in a smooth, graded fashion, and we *experience* our eye movements as a continuous experience. Physiologically speaking, eye movements result from the summation of multiple neurons firing and many very small increments of movement. At the individual neuronal level they are physiologically categorical, but experientially they are continuous.

These examples demonstrate that many events in nature can be described as categorical at a microscopic level and graded as the result of summing at a macroscopic level. Keeping the two models distinct allows for in-depth study and avoids oversimplification. Thus, keeping the two models distinct should not be misinterpreted as implying that one is better, more powerful, or more reflective of nature.

In sum, each model better explains certain properties of a system at a specific level of analysis. Whether these distinctions describe actual differences that occur in nature, that is, distinctions that could not be perceived in any other way, is a debate that leads back to Plato's cave. I conclude that the distinctions between the categorical and the probabilistic remain useful at this point in human history. Whether they will continue to have conceptual value in the future is unclear, even if Kant is right in claiming that the way humans perceive and conceive of the world is shaped by the way the human brain is constructed. The next chapter suggests that there is at least one other conceptual model of causality, the emergent approach, which has been developed over the past hundred years. The fact that each of these models became explicitly recognized at certain points in human history supports the notion that new conceptual models will develop in the future and that these could unite or replace current approaches.

CRITERIA FOR CHOOSING THE APPROPRIATE MODEL

If the categorical and probabilistic models are both useful, there should be criteria that guide the choice of which model to apply in a given situation. Two have been already been indirectly discussed. First, statisticians have developed rules that dictate the circumstances in which a specific statistical technique should be used; almost universally, different statistical methods are appropriate for analyzing data that are categorical or dimensional. This is not circular reasoning because the appropriate choice maximizes the "power" or ability to provide an answer. If the question can be stated in categorical terms, then techniques that give absolute "no/yes" answers are utilized. If not, probabilistic techniques are used. However, categorical methods still have a probabilistic component since absolute certainty is not possible and because the number of episodes or individuals influences the strength of the likelihood. The result is that the correct statistical method maximizes the degree of certainty with which one can show that a relationship exists, and, in the appropriate circumstances, it quantifies the likelihood of causality.

Utility is a second criterion that is sometimes appropriate to invoke. As already noted, meteorologists distinguish between tropical storms (sustained wind speed is between thirty-nine and seventy-five miles per hour) and hurricanes (sustained wind speed is greater than seventy-five miles per hour), and, more recently, hurricanes have been further subdivided into five subcategories. These distinctions are clearly artificial because wind speed is a graded function, and there is no clear difference between storms that have mean wind speeds of seventy-four miles per hour and those with seventy-six-mile-per-hour winds. However, these categorical distinctions serve communication (people listen more attentively when the word hurricane is being used) and disaster-preparedness functions (different, more drastic actions are taken for a level 5 storm than for a level 2 storm). On the other hand, if one were trying to study whether the amount of damage done by a storm correlated with the amount of energy in the storm, then equations should be used that examine wind speed and other variables (such as storm diameter) as continuous, dimensional constructs.

Not surprisingly, even in science the choice of model may change over time if the type or amount of information changes. One hundred and fifty years ago, Gregor Mendel demonstrated that eight characteristics of peas are inherited as individual elements, later named genes, from each parent and that this inheritance follows a pattern in which the inherited elements act in one of two ways. Those that are "dominant" induce the trait no matter what the other element is; those that are "recessive" will be expressed only if the other gene is also inherited. Mendel's work fell into obscurity but was "rediscovered" around 1900, and this categorical/digital description guided much of genetic research for one hundred years. With the sequencing of the human genome in 2001 and the subsequent ability to sequence whole genomes of many individuals, the limitations of Mendel's categorical/digital model have become evident. In fact, most common illnesses in humans that run in families appear to result from many genes, each of which explains only a small portion of the inheritance. This does not mean that Mendel was wrong, only that genetics does not follow a simple categorical pattern in many if not most situations. Scientists are working to unravel this complexity. The solution will clearly incorporate Mendel's insights, since they do explain many phenomena, but it is likely that mechanisms will be discovered that depend upon interactions among multiple genetic elements and result in graded, dimensional traits. It is also possible that mechanisms will be discovered that appear to result in categorical gene expression but actually depend upon nonlinear biological interactions, the topic of the next chapter. The choice of which model to utilize will depend upon which model better predicts the function of specific genes and sets of genes in the natural world. In the end, then, the usefulness criterion becomes very powerful.

SUMMARY

This chapter has explored the world of complex, interacting, graded, and multiple causes. Some causes act individually but do so in a graded fashion. Others act as systems in which many variables—gene products, signal transduction pathways, cell systems, neuronal networks, organs—interact in predictable ways.

It is the word "predictable" that is crucial here. While chance may seem to be the explanation, mathematical analysis demonstrates that probabilistic relationships are at work. It is the probabilistic nature of the relationship that allows predictions to be made and that is the strength of this approach. However, the inability to make absolute statements is a source of frustration to many.

The next chapter considers phenomena that seem to combine aspects of the two models discussed thus far—phenomena that have both the all-or-none characteristic of the categorical model and the graded characteristic of the probabilistic model.

5

A THIRD MODEL OF CAUSALITY
The Emergent, Nonlinear Approach

What we are witnessing . . . is a change of paradigm in attempts
to understand our world as we realize that the laws of the whole
cannot be deduced by digging deeper into the details.

—Tamas Vicsek

These laws, applying equally well to the cell and the ecosystem,
demonstrate how unavoidable nature's laws are and how deeply
self-organization shapes the world around us.

—A.-L. Barabási

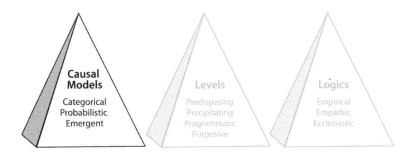

In the late 1940s, the MIT professor Norbert Wiener proposed a
new approach to the study of systems: he analyzed them as functioning wholes made up of multiple interacting modules. He named the
discipline cybernetics and contrasted its approach to the prevailing
scientific model's focus on individual interactions. Inherent in Wiener's proposal was the idea that the sum of the parts of a system is
greater than their individual contributions. This idea seemed quite
radical at the time, in part because Wiener had difficulty finding convincing real-world examples that he could use to convince skeptics,
but it influenced the designers of the first computer systems and can

be seen as a precursor to the relatively new disciplines of systems biology and network theory.

In fact, these ideas were not really new. As Wiener himself acknowledged, the idea that the whole is more than the sum of its parts had been prominent in Eastern thought for several thousand years. This idea had also been central to the work of the physiologist Claude Bernard, the mathematician Henri Poincaré, and the philosopher George Henry Lewes, all of whom worked in the mid- and late nineteenth century. For example, Bernard's theory of homeostasis posited that organisms have multiple systems whose interactions maintain a steady internal milieu in the face of a changing external environment, and Poincaré developed mathematical tools to analyze relationships between variables that seemed to vary irregularly, that is, nonlinearly.

Wiener's emphasis on studying systems from the point of view of integrated and interrelated wholes rather than as a conglomerate of standalone, self-contained units contains within it the concept of *emergence*, the idea that systems contain organizational elements that cannot be understood solely by examining the constituents individually. These concepts of systems, nonlinearity, and emergence and their relevance to causality are the focus of this chapter.

Over the past half century, several conceptual models have built upon the tenuous skeleton of Wiener's ideas and extended them to the physical, biological, and social sciences. Topics as disparate as the functioning of organisms, economic systems, nonhuman animal and human societies, accidents, historical movements, disease epidemics, human consciousness, and even the universe and life itself are being explored with methods that utilize the concepts of nonlinearity, emergence, and systems. These methods go by names such as chaos theory, complexity theory, self-organizing systems, and network theory. While it is too early to be sure that they share a coherent, productive, and predictive set of techniques, the broad approach that is common to them represents a third conceptual model of causality that describes aspects of causal relationships not captured by categorical or probabilistic reasoning. The following paragraphs develop this idea by reviewing the central theses of several of these models, with an emphasis on identifying characteristics common to their approach to causality.

CHAOS THEORY

Chaos theory utilizes a set of mathematical techniques to identify predictable patterns of behavior within seemingly unorganized, repetitive activities and to demonstrate that there are probabilistic boundaries within which these phenomena take place. The use of the word "chaos" is somewhat ironic, since the theory seeks to identify both regularity and predictability in phenomena or observations that appear to be random, but the name is catchy and does emphasize the idea that seemingly unordered, chaotic phenomena are the activities the theory seeks to explain.

Three central ideas underlie chaos theory and shape its contribution to this third model of causality: *the importance of initial conditions, nonlinearity, and emergent interactionism.* The first concept, the claim that initial conditions significantly influence the course and outcome of an activity, is supported by the observation that subtle variations at the beginning of many activities lead to profoundly different results. A mundane example is that a stone rolled down a hill several times will follow different paths no matter how much effort is put into starting the activity exactly the same because it is impossible to begin each roll in exactly the same manner and because each stone is different. The important role of initial conditions places major limits on predictability because it is very difficult to describe or identify every nuance of the initial circumstances of an event. This in turn places limitations on the identification of initial causes or the specifications of the beginnings of a causal chain.

The second thesis of chaos theory is that many activities that appear random, patternless, and chaotic can be described by non-linear mathematical techniques. The next section of this chapter expands upon the concept of nonlinearity and contrasts it with the dichotomous relationships of the categorical model and the graded aspects of dimensional reasoning.

The third concept of chaos theory is that an analysis of events at the systems level can reveal patterns of regularity not evident when the focus is on individual elements of the system. This derives in large part from the high frequency of nonlinear relationships. These patterns often seem to emerge from random activity, but chaos theory suggests that it is the focus on individual elements rather the whole that has prevented the pattern from being recognized.

Proponents of chaos theory claim that the repeated demonstration of nonlinear regularity in systems as disparate as heart arrhythmias and weather suggests that there is an underlying, shared causal principle. So far, though, chaos theory has provided only descriptive tools that identify nonlinear relationships in systems, not insights into how events are generated. That is, it has yet to contribute to an understanding of mechanism. However, one could make the same statement about Newton's laws or the laws of thermodynamics, so chaos theory's description of regularity and specification of rules by which systems work are contributions in themselves. It is the demonstration that initial conditions and nonlinearity add to our understanding of processes as different as nerve synapses, ecological niches, and weather and the fact that categorical and dimensional/probabilistic causality do not fully explain their functions that supports the idea of a third model of causality.

NONLINEAR DYNAMICS

Many events that seem to occur in an all-or-none fashion are found, on closer examination, to result from an accumulation of changes that have occurred over a longer period of time. Familiar examples include earthquakes, the freezing of pure water at thirty-two degrees Fahrenheit (zero degrees centigrade), and the boiling of water at 212 degrees Fahrenheit (one hundred degrees centigrade). Other examples of events that follow this pattern are the firing of nerves, the development of "superconduction" (the ability to conduct an electrical current with no resistance) in some materials when they are cooled to very low temperatures, and the change from being an insulator to being a conductor that occurs in other materials when they are cooled.

Each of these events appears to happen suddenly and in the all-or-none manner of the categorical or binary causal model. However, each follows the "accumulation" of a graded change over time—a change in temperature/energy for the formation of ice, steam, and the superconducting state; a buildup of pressure resulting from the collision of tectonic plates that leads to earthquakes; a change in ion concentration and electrical charge that occurs before a nerve fires—a characteristic of dimensional, probabilistic cause. It is this combination of a gradual accumulation resulting in sudden change that is the source of the label "nonlinear."

The specific change undergone by that material or entity in each of these examples is dictated by its makeup at both the subatomic and macroscopic levels. That is, the cause of these sudden transitions, whether it is the change from a conducting state to an insulating state, from a solid to a liquid, or from a resting state to a depolarized state, lies within the structure of that specific material—an inherent property of that substance or structure. When ice forms, when a liquid turns to a gas (steam), or when a nonconducting substance turns into an electrical conductor, the suddenness of change in the macroscopic makeup of the substance is accounted for by an "instantaneous" rearrangement in the quantum, molecular, and macromolecular organization of the constituent units of that substance. Earthquakes are caused by the sudden movement of one or several tectonic plates and the energy released as the result of this movement. The energy buildup occurred because the plates are being pushed into each other; the suddenness of the event and the large amount of energy released results from the tremendous amount of energy needed to overcome the forces of mass and gravity that have kept the plates in place.

The study of phase transitions and the nonlinear mechanics that best describe them has begun to provide insights into the functioning of complex systems, and this has invigorated Norbert Wiener's rather vague notion of cybernetics. Several generalizations can now be made about the characteristics of nonlinearity that provide a springboard for both defining and characterizing nonlinear causality.

First, nonlinear change occurs in systems that have a large number of elements. One or two water molecules would not form ice, nor would a system made up of only two tectonic plates generate an earthquake. The presence of a large numbers of elements increases the number of potential interactions and increases the probability that an uncommon or unanticipated outcome will occur.

Thus, limited predictability is a second general aspect of nonlinear systems. Per Bak and Kan Chen, who pioneered the study of nonlinear mechanisms in earthquakes and identified several of these general characteristics, cite the nonlinear aspects of weather patterns, avalanches, and traffic jams as reasons for the limited success in predicting them, although this ability has improved as the identification and quantification of contributing elements has improved. However, their own field of earthquake prediction has still not yielded a model accurate enough to make useful predictions.

Another aspect of systems that limits predictability is the considerable role initial conditions play in determining subsequent events in many systems. Because of this, accurate predictions cannot be made in many systems until the specific circumstances of the initiating event are known, that is, until the sequence has actually started. This is a restatement of the sensitivity to initial circumstances characteristic of chaotic systems.

On the other hand, many systems have "built-in" aspects that constrain the number of interactions or redirect subsequent interactions toward the "intended" goal; these characteristics counterbalance the limited predictability imposed by initial conditions. For example, many systems are constructed such that interactions occur in prespecified sequences, in which a specific outcome triggers the next event in the sequence. If the event initiating a sequence is also constrained by built-in system elements, predictability is further increased and the unpredictability of initial conditions lessened.

A third generalization about nonlinear systems is that outliers, that is, unexpected events occurring far from the mean, are *more* likely to occur. As a result, dramatic change is more common in nonlinear than in dimensional causal relationships. This further lessens the ability to make predictions in nonlinear systems. Said another way, linear causal models associated with dimensional, probabilistic relationships describe predictable relationships between variables and the mean of their distribution. In nonlinear models, the relationship between the mean and an individual element is more variable. Thus predictions based on knowledge of a small number of elements of nonlinear systems are less accurate than those based on a small number of elements of categorical or probabilistic causal models.

Because rare or outlier events are more difficult to identify or observe and because they are more common in nonlinear systems, it is harder to identify elements present around the time that the nonlinear causal chains begin. This makes it more difficult to design tests that replicate the conditions that existed at a given time of interest when studying causal modeling. Whether the more complex mathematical modeling required to characterize nonlinear systems is less intuitive to humans than categorical or probabilistic modeling is unknown, but this idea bears testing because it has implications for educating the public.

An example of this challenge is the power-law relationships characteristic of some nonlinear systems. In comparison to the Gaussian distribution, the frequency distribution predicted by a power law quickly reaches a maximum point and then declines more slowly because the exponent in power-law relationships is a fraction. (In the example $2^3 = 8$ [$2 \times 2 \times 2 = 8$], the 3 digit is the exponent.) This describes the fact that a change in one variable leads to disproportionate change in the other. For example, the relationship between the size of an organism and its metabolic rate follows a power law that, for most organisms, is described by the exponent $\frac{3}{4}$; this means that an increase in size of three units is accompanied by (or requires) an increase in the metabolic rate of four units. Since there is a limit on how much energy a system can generate (its maximum metabolic rate), there is a limit on how large any given organism can become, and that limit is driven by the denominator (the bottom number in the fraction), the metabolic rate.

A fourth characteristic of nonlinear causality that is illustrated by the power-law relationship is that some changes that precipitate an event appear to be quite small. The formation of ice and the development of superconductor status, for example, seem to occur after small changes in temperature. Likewise, nerves depolarize or "fire" after the application of a small amount of electricity or a small change in the chemical makeup of the internal milieu of the cell. However, this sudden occurrence belies the fact that significant energy has been gradually removed from or added to the material before the final triggering change has occurred.

Bak and Chen coined the phrase *self-organized criticality* to label this critical point of sudden change and to emphasize that the point of change in any given system depends on specific structural characteristics (the self-organized aspect) of the elements that make up that system or material. Sudden change also occurs in quantum-level phenomena. Referred to as *singularities*, they reflect instantaneous changes in mathematical properties that occur when a subatomic particle suddenly acts like a wave or becomes infinite, when the energy contained in a relatively small number of molecules is released in an atomic bomb, or during the conditions at the beginning of the current universe, when an infinitely small point of energy began expanding into the mass that became stars, planets, and everything else.

One of the characteristics of the binary categorical causality described in chapter 2 is the distinctiveness between what went before and what came after, that is, between cause and effect. This sudden change in status is also characteristic of nonlinear causality, whether it is labeled self-organized criticality, a singularity, or the peak of a power-law relationship. However, in the categorical approach change has not usually accumulated prior to the observed change, while in the nonlinear model the opposite is true. Many historical events, fads, and disease epidemics follow this pattern of appearing to occur suddenly after a period of gradual and often unrecognized change or accumulation. Because the event occurred suddenly, a proximal preceding event is frequently identified as its primary genesis, a mistaken application of categorical causality. The electricity blackouts of the past two decades in the United States are examples. The failure of a single, often small substation is frequently initially cited as the precipitating culprit, but later analysis demonstrates that that failure alone could not have caused the whole system to fail. Indeed, a specific cause of most widespread power failures is never found, and the blame is ultimately laid at the foot of the "system." Thus, when examination of preceding events reveals a gradual accumulation of multiple changes, or if "built-in" or preordained elements are identified as having a causal role, a nonlinear, more complex explanation should be sought or at least considered.

There are also similarities in appearance between nonlinear and dimensional causal dynamics. In both, the outcomes result from the interaction of several or many events. However, in the dimensional model these interactions follow a gradual, additive (or multiplicative) pattern rather than the less regular power-law pattern of nonlinear systems.

As noted previously, a primary purpose of recognizing which model is operating is to alert the seeker of causality to the types of relationships that should be sought among the variables of interest. This raises a conundrum. Does the model explain why a specific causal relationship exists, or is it merely a description of what is possible in the universe we experience? The argument made here is that it does both—these models reflect basic properties of the universe and thus describe broad categories of interactions, but they also impart structure and limitations to the search for causal interactions and can be used to demonstrate the existence of such relationships.

One other issue that places limits on predictability in nonlinear systems as well as in binary and dimensional interactions is the crucial role that unique aspects of individual constituents sometimes have on ultimate outcome. Superconductivity occurs in very limited circumstances because the atomic structure needed to sustain this phenomenon is dependent on unique characteristics of certain rare earth metals and their interactions with other compounds. Some generalizations to other compounds have been possible once an understanding was gained of the initially discovered compound, but the ability to predict the specific makeup of other superconducting compounds is still limited. Likewise, self-organized criticality is usually identified only after it has occurred on multiple occasions. Perhaps a better understanding of nonlinear dynamics will lessen this limitation in the future, but dependence on innate characteristics of specific elements or situations will always lend an aspect of unpredictability to nonlinear causality as well as to situations in which categorical and dimensional cause are applicable.

A final aspect of nonlinear causality that bears mentioning is that it provides an approach that combines "top-down" and "bottom-up" approaches. The top-down approach begins with a systemwide, big-picture view and identifies interactions at that macro level. The bottom-up approach, on the other hand, starts with the smallest elements and builds a causal explanation based on the interactions at the micro level. Among the bottom-up elements of nonlinear causality that have been mentioned are the unique, inherent, pre-existing aspects of particle physics, single molecules, and individual units or modules of body organs. Top-down features include the emergent qualities that are expressed in the power-law relationship and the interrelationships existing among the structural elements that cannot be captured in a simple catalogue of a system's elements. It is the recognition of the existence and importance of these top-down elements that distinguishes Aristotle's third level of program-matic causality from predisposing and precipitating causes. Elements of predisposition and precipitation are clearly not excluded, but this level of analysis adds something that neither represents, something that chaos theory, systems biology, and network theory are beginning to explicate.

The benefits of combining the bottom-up and top-down approaches were nicely highlighted by the Nobel Prize–winning physicist P. W. Anderson in a 1972 article whose thesis he summa-rized as follows:

> The ability to reduce everything to simple fundamental laws does not imply the ability to start from those laws and recon-struct the universe. . . . Instead, at each level of complexity, entirely new properties appear, and the understanding of the new behaviors requires research which . . . is as fundamental in its nature as any other.

GENERALIZING FROM NONBIOLOGIC TO BIOLOGIC SYSTEMS

The discussion of nonlinear systems has thus far focused primarily on nonbiologic examples. Even though system-level processes such as homeostasis have been studied by biologists for more than a cen-tury, the operation and genesis of system-level processes has engen-dered renewed interest in the last decade. For example, in *Foundations of Systems Biology* (2001), Hiroaki Kitano and coauthors describe a number of applications of a systems approach to biological sys-tems. In the opening chapter, Kitano notes: "There are interesting analogies between biological systems and engineering systems. Both systems are designed incrementally through some sort of evolutionary process, and are generally suboptimal for the given task. They also exhibit increased complexity to attain a higher level of robustness and stability."

Kitano goes on to identify four general elements of "engineering systems" that enhance robustness and stability and suggests that they might also be intrinsic aspects of biological systems:

(1) *System controls*: Most biologic systems include *feedforward* controls, in which a given stimulus sets off a series of connected steps that occur in a predetermined sequence. The blood-clotting system is an example. When a breach in a blood vessel occurs, molecules are released that interact. Their interaction then triggers another step. This cascade of very specific biological steps continues until a clot is formed that stops the bleeding (or fails to do so, and the organism dies of blood loss). However, *feedback* mechanisms, by which systems provide information to prior steps in a sequence and thereby adjust output to conditions detected by an on-line monitor, are often more important for the long-term maintenance of a system. Any number of examples can be given; the control of blood sugar by insulin secretion is a well-known one.

(2) *Redundancy*: Many biologic systems have duplicated processes or pathways for reaching a desired outcome. The bilateral symmetry (one side of the organism is a mirror image of the other) of many organisms and the resulting duplication of organs is an example at the organ level. Familiar exemplars of this are the two kidneys, lungs, and eyes that are characteristic of many mammals.

(3) *Structural stability*: Some systems have more than one path or process for achieving a given goal. This increases the likelihood that the system will achieve its goals under varying environments or circumstances. If one subsystem is damaged, or if environmental circumstances change significantly, the presence of multiple mechanisms or pathways thereby increases the likelihood that varying circumstances can be adapted to or overcome. For example, the human immune system has both a preexisting, "on-line" component and an induced antibody component to respond to foreign "invaders." Some organisms have multiple pathways and utilize them in different

environmental circumstances. Redundant systems accomplish the same end but do so with more than one similar or equivalent pathway/organ. (One criticism of the Chernobyl reactor was that it did not have multiple pathways to guard against overheating, so that when the single system failed, disaster was inevitable.)

(4) *Modular design:* The encapsulation or clustering of individual subsystem elements and their separation from other subsystems and elements appears to have several benefits. It can limit damage when problems occur by lessening the likelihood that damage will spread to the whole system. Modular design also allows for easier replacement if damage occurs and fosters the development of improved processes by maximizing the impact of change within a module. By isolating crucial elements of a system, modularity provides a mechanism for the development of "rate-limiting" or controlling steps through which many disparate inputs funnel. This, in turn, provides mechanisms to maximize output control. For example, the liver serves as a toxin-clearance center because it contains cells that metabolize or break down compounds through several different mechanisms. This function is maximized because the body's blood vessels are organized to assure that a large percentage of blood flows through the liver on each cycle around the body. Thus, the liver is a detoxification module that consists of several unique and distinct approaches to the task of detoxification. Similarly, the production of insulin by the pancreas, a crucial step in the maintenance of a continuous supply of the body's energy source, glucose, is under the control of a series of mechanisms by which body sensors detect the availability of fuel (blood glucose) and control the amount of insulin that is released. The "sensors" and the cells that make insulin are clustered together. This proximity presumably maximizes the fine control needed to access energy very quickly and simultaneously avoids the toxicities associated with too much or too little glucose.

The hope of systems biology is that the demonstration of such commonalities across species, genera, and kingdoms, as well as across biological abstractions such as niches and ecological systems, not only describes nature but will lead to the discovery of mechanisms that underlie such systemwide activities. Two fields in which some progress is being made in this regard are genetics and organism development. For example, a recent study by A.-C. Gavin of all the proteins within yeast cells demonstrates that the molecular "machinery" within protein-producing cells cluster into 257 unique groupings, an example of modules at the level of cellular processing. The promise of systems biology is that the study of these modules, both how they fit together and how they become impaired, will be furthered by studying how the modules interact with one another. In another study, C. Pal and coworkers used computer modeling to repeatedly "expose" ancestor organisms to specific environmental conditions and found that the organisms' current genetic composition could be predicted with 80 percent accuracy. Such studies raise the possibility of identifying system-level ("top-down") principles of biology and advancing knowledge about how they operate at the level of speciation evolution.

Systems biology, then, seeks to identify organizational features that meet criteria for programmatic causality. It relies on data that are both dimensional and categorical in nature, but it seeks to identify broader mechanisms that operate across multiple biological systems rather than within single systems. This combination of analyses at multiple levels is reminiscent of the Aristotelian notion of causality.

SELF-ORGANIZATION

A somewhat different approach to network theory in biological systems is taken by Scott Camazine and coauthors, in their book *Self-Organization in Biological Systems*. They begin by defining *self-organization* as "a process in which a pattern at the global level of a system emerges solely from numerous interactions among the lower-level components of the system. Moreover, the rules specifying interactions among the system's components are executed using only local information, without reference to the global pattern."

While acknowledging commonalities between the laws of physics and the patterns of simplicity underlying complex biological activity, Camazine et al. emphasize two major differences between emergent behaviors in biological and physical systems. One is that the presence of genetic controls in biological systems provides a set of commands that "finely tune the rules of interaction," a level of control that is lacking in many physical systems. Second, biological systems tend to be more complex than physical systems because they are made up of a significantly larger number of interacting modules or elements.

Camazine goes on to identify five characteristics that exemplify self-organizing systems:

1. Positive and negative feedback
2. Interactions among the "lower-level" components
3. Tunable emergent features
4. Stigmergy (the guiding of an activity by previous activity rather than by an external program, for example, the shaping of a nest or hive by the twigs that have already been placed)
5. Decentralized control

This list shares with Kitano's the principles of inherent control systems that adjust and adapt to changing internal and external environments (features 1, 2, and 5). It adds the programmatic feature of emergent properties and mechanisms by which chance elements in the environment involved in crucial functions (for example, the shape of already-used twigs) can be adapted to or directly incorporated into its output. It also suggests that systems biology will need to clarify the mechanisms of emergence and stigmergy if it is to become a discipline of its own.

NETWORK THEORY

Another approach to the question of how individual units interact to form systems emerged from research in the 1960s by the psychologist Stanley Milgram and the sociologist Mark Granovetter. They demonstrated that the interactions among humans are not isolated to single pairs of individuals and do not occur randomly. Rather, individuals tend to interact in concentrated groups that have come to be called "small worlds." These groups connect occasionally with other groups by what are called "weak links." What is surprising about their research is the finding that these "weak links" play important roles in people's lives. For example, Granovetter found that weak links such as minor acquaintances are common sources of finding a job, even though the vast majority of interactions people have are in the small world of friends and relatives.

By the 1990s, Steven Strogatz and Duncan Watts reported finding similar small-world and weak-link relationships in the large electric power grids so crucial to present-day life and in the neural networks of the nematode *C. elegans*, a widely studied organism whose nervous system had been completely mapped by the early 1990s. They suggested that the similarity of organizational features in such widely different constructs as neural networks, power grids, and human interactions raised the possibility that a hitherto unstudied level of organization exists in nature.

The development of the World Wide Web provided Albert-Laszlo Barabási and colleagues with new methods and models for identifying mechanisms by which such systems operate. Like Granovetter and Milgram, they found that a small number of websites ("hubs") were widely linked to other websites, while most web sites ("spokes") had very few links. In addition, they found that the distribution of linkages among all websites follows the power-law distribution. Barabási and colleagues have subsequently studied a number of economic, political, and biological systems and demonstrated that many can be described in a similar fashion—what they call "scale free." By this they mean that a few, large hubs are dominant and therefore have prominent roles in determining the operations and outcomes of the entire system. For them, the finding that a few hubs are very widely connected reveals mechanisms by which a very small number of hubs play major roles in large disseminated systems.

The exponential growth of the World Wide Web during a brief period in the 1990s and early 2000s also provided Barabási and colleagues with a model system for studying the effects of *time and timing*, variables that have been overlooked in causal modeling. In studying how the addition of new links affected the development of the Web, they concluded that early sites have a distinct survival advantage because it was easier for them to become widely connected when fewer sites existed. Becoming widely connected early in the development process increases the likelihood that a site will become even more connected over time. This is why fewer sites and links that form later in the developmental process become hubs. This finding is reminiscent of the claim in chaos theory that early conditions have a disproportionate role in shaping subsequent events. However, Barabási and colleagues have gone on to show that subsequent events do play a shaping role in the structure and that the advantage gained by being an early network participant can be overcome if the newcomers have distinct advantages over existing hubs. The emergence of Google as a search engine several years after the founding of others is an example of the latter point.

One implication of the dependence of networks on relatively few, highly connected hubs is that the removal or destruction of a few crucial hubs can have a devastating effect on the functioning of the whole system. As the failure of the power grid has demonstrated, widespread failure can result from the shutdown of a small number of central hubs (modules), especially if the network is designed in such a way that alternative systems (redundancies, duplicates, or alternative paths) cannot be called upon to take over. While this is not surprising (the blockage of a single artery in the heart can cause the heart to stop functioning as a pump and lead to the death of the organism within minutes), the identification of this as a causal/descriptive principle helps explain why dramatic changes can occur over short periods of time when small changes have been accumulating over a longer period of time. One example is the punctuated theory of evolution proposed by Eldridge and Gould, in which sudden, rapid evolutionary change occurs after long periods of the more gradual accumulation of change suggested by Darwinian theory. Another example is the "great-person" or single-event approach to history, in which dramatic changes are attributed to single events or individuals after a gradual accumulation of many events and the work of many people over a longer period of time. Both of these are examples of causal models that combine the insights of the emergent, nonlinear model of systems theory with those of the dimensional and categorical models.

SUMMARY

A new model of causality has emerged in over the past century. The three fields of inquiry reviewed in this chapter, chaos theory, systems biology, and network theory, provide convergent evidence that this model of emergent or nonlinear cause is manifest in such diverse constructs as weather, protein regulation, disease spread, and group behavior. Although each of these fields has emphasized different elements (and each is still a construct under development), the features they share suggest that this model of causality is characterized by:

1. Exaggerated influence of early or initial conditions
2. Modularity, that is, groupings of elements into functioning systems within which the elements primarily interact with one another (e.g., small worlds, modules, and hubs)
3. Connections among its parts and structures that enhance both the ability to withstand changes and to utilize alternative approaches if that would enhance the survival of the system (e.g., feedforward and feedback systems, redundancy, alternative paths with similar outcomes)
4. Scale-free or power-law relationships among elements

As noted earlier, the concept of systems is not new and has been expressed in many forms. Eastern thought has long emphasized the interrelatedness of nature's many elements. Aristotle's third level of causality, both originally and as reformulated in chapter 1, contains within it the idea that causality can emerge from the interactions among a number of elements that make up a system. Bernard's idea of homeostasis invoked the existence of physiologic systems that enhance stability in the face of changing circumstances. The existence of many subdisciplines of the physical (from subatomic to macroscopic) and biological (molecular to ecology) sciences also suggests that different levels of analysis exist but that there must be interactions among these levels. Whether this is truly a new model of causality or merely an emphasis on certain aspects of existing models of reasoning can still be argued, but if the new tools being developed by system-focused disciplines improve the analysis of cause and provide an understanding of nature that cannot be gained through other models, then the revolutionary claim that this is indeed a new conceptual model of causal reasoning will be supported.

6

EMPIRICAL: THE PHYSICAL SCIENCES

Science has managed to discover a great deal about the world and how it works, but it is a thoroughly human enterprise, messy, fallible, and fumbling; and rather than using a uniquely rational method unavailable to other inquirers, it is continuous with the most ordinary of empirical inquiry.

—Susan Haack

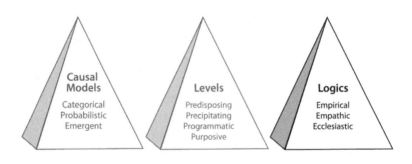

Because many of the events and objects studied in the physical sciences can be readily observed, manipulated, and measured, chemistry and physics should contain some of the best-established examples of causal reasoning. Perhaps this is the reason that scientists such as Galileo and Einstein have influenced not only our current understanding of how the universe works but also how reality and causality are conceptualized. This chapter reviews three concepts developed during the twentieth century by physicists, chemists, and mathematicians—relativity theory, quantum mechanics, and the incompleteness theorem—that have significantly influenced the current conceptualization of causality and must be considered in any general approach to the subject. It also explores how a series of discoveries in the physical science of geology led to the development of plate tectonic theory and how, after a long period of skepticism about its validity, this theory became widely accepted as the primary causal explanation for geologic phenomena as diverse as continents, earthquakes, and mountains.

This chapter will also review in more depth three concepts already mentioned: First, there are limits to the knowledge one can gain no matter what techniques are used. Second, more than one approach or vantage point is sometimes necessary to identify causal influences with as much accuracy as possible. And third, in many complex physical systems, causality results from the interaction of multiple elements rather than from the interaction of two individual events.

Both quantum mechanics and relativity theory emerged at the beginning of the twentieth century from the attempt by the German physicist Max Planck to explain an anomaly he noted when he was investigating black-body radiation, a phenomenon in which heat (energy) is generated when two thin metal plates are placed very close together. The anomaly Planck puzzled over was the demonstration that more energy was generated than predicted by the then current theory that light and energy were propagated as waves. Planck showed mathematically that the "extra" heat could be explained if energy were contained in indivisible energy packets, or quanta, rather than waves. This discovery launched the field of quantum mechanics. It was refined over the next half century by Planck, Niels Bohr, and others, and it remains an active contributor to our understanding of the cosmos. While the details of quantum physics are not germane for the discussion here, several of its discoveries have had a significant influence on current ideas about causality and will be enumerated in this chapter.

INDETERMINACY: ABSOLUTE KNOWLEDGE IN THE PHYSICAL SCIENCES IS NOT POSSIBLE

As noted in chapter 1, the physicist Werner Heisenberg stated in 1929 what has become known as Heisenberg's uncertainty principle:

It is not possible to know both the position and momentum (speed) of a sub-atomic particle. The measurement of either one can only be accomplished by affecting and thereby altering the other.

The intriguing implication of this principle is that, at least in the sub-atomic world of quantum physics, it is not possible to know "everything" about an event or an object, because measuring one property unavoidably alters others. As a result, fully accurate description is not possible and can only be estimated.

It seems reasonable to ask whether this applies only at the sub-atomic level or extends to the macroscopic world of our everyday experience. This remains a topic of some disagreement, but the identification of a limit to absolutely accurate knowledge remains powerful and has had a profound impact on the modern conceptualization of causality.

Two years after Heisenberg published his principle, Kurt Gödel demonstrated that there are also limits to knowledge in the field of mathematics. Gödel was examining the failed attempt of the British philosopher/mathematicians Bertrand Russell and Alfred North Whitehead's *Principia Mathematica* (which had been published between 1910–1913) to derive all mathematical logic from a set of basic principles, a goal of mathematicians since the ancient Greeks. Gödel provided a mathematical proof, which came to be known as the incompleteness theorem, demonstrating:

It is impossible to describe a mathematical system in which all formulae can be derived from a given set of laws or assumptions.

What Heisenberg's uncertainty principle and Gödel's incompleteness theorem identify, then, are limitations to absolute knowledge in the physical and mathematical sciences. This claim that the goal of *complete* knowledge is unobtainable is relevant to all scientific disciplines, since matter is the basic material of the universe and mathematics a basic tool for describing the relationships of the building blocks of matter to one another. For our purposes, it is a claim as strong as that made by David Hume in the late eighteenth century that there are limits on causal knowledge.

It is important to note that this limitation in the ability to measure nature does not state or imply that exact information is unobtainable or that it is impossible to learn a great deal about a subject. The extraordinary accuracy required to land a probe on the asteroid Eros and the ability to measure the energy of individual subatomic particles are examples of just how precise mathematical calculations and predictions deriving from them can be. What the uncertainty principle and incompleteness theorem do tell us, though, is that, in any given situation or given system, there will always be facts that are unknowable.

That there are limits to knowledge shouldn't be surprising. One of the great discoveries of fifteenth-century physics was Kepler's recognition that the planets move in elliptical orbits around the sun. He showed that simple equations could both describe the motion of the planets and make accurate predictions of their location in the foreseeable future. However, these predictions, like those derived from Newton's laws of gravity, relied on the assumption that gravity is operating on two bodies—the sun and the planet of interest. Even today, it is not possible to describe mathematically all the forces involved in the interaction of *three* bodies, much less among four, ten, or a thousand "particles" or heavenly bodies. It is probably a mistake to state that this will never be possible, but it is certainly extremely unlikely in the near future and may never happen.

Need we be paralyzed by the existence of these limitations to our ability to gain fully accurate knowledge? Absolutely not. What the many engineering and scientific successes of this and former eras (some of which are described in the following chapters) tell us is that we can make very accurate causal predictions in spite of our inability to have *absolute* knowledge. As discussed in chapter 3, these calculations have margins of error, but their precision is extraordinary.

One important and somewhat ironic lesson that has been learned from the uncertainty principle, the incompleteness theorem, and the limited ability to predict the interactions of multiple objects on one another is that the recognition of a limit to our ability to know can be a spur to the development of methods for increasing the accuracy of what *is* measurable and knowable, for quantifying what is unknown and unknowable, and for identifying areas in which advances in knowledge are more likely. Perhaps like the carpet makers who incorporate a mistake into every rug as an acknowledgment of their imperfection as human beings, those who acknowledge the existence of limitations are freer to search for what is knowable.

In applying these ideas to the search for causality, it appears that Hume and Aristotle were both correct. Hume was right to say that we are limited in our search for causal certainty and also right to note that there are actions we can take to improve our accuracy. Aristotle was right to claim that we can identify causes and to propose that differences in accuracy might characterize the different levels of causality.

Clearly, knowing the degree of uncertainty present in any given circumstance is a powerful piece of information, and the development of statistical methods for describing and estimating the degree of uncertainty has been an important advance in many areas of applied and exploratory science. If certainty is close to 100 percent, then there is the possibility that causality can be established to a very high degree of likelihood. If the level of uncertainty is high, then it is likely that other or multiple causes are present. Clearly, absolute certainty about causality is not possible in those disciplines in which mathematical relationships and physical forces are operating. As we will see in chapter 11, medicine and law have incorporated the inevitability of uncertainty into everyday practice by incorporating estimates of this uncertainty into their everyday functioning.

ABSOLUTES IN AN INDETERMINATE WORLD: THE DUALITY OF LIGHT AND MATTER

Isaac Newton made many important contributions to scientific knowledge. Among the most influential were his lectures on optics, delivered between 1669 and 1671 and published in 1729. In them he demonstrated, through an elegant set of experiments using prisms, that many of the properties of light could be explained if light consisted of a set of waves. Color, for example, is a property of different wavelengths. The Newtonian model was supported by nineteenth-century experiments in which the interference patterns observed when light passed through a grating of very narrow slits were also best explained by the hypothesis that light had wavelike properties.

However, the demonstration by Albert Michelson and Edward Morley in the late nineteenth century that the speed of light was the same in all directions led Einstein to propose that light also had particle-like properties. This seeming contradiction, that light had the properties of *both* a wave and a particle, suggested to Einstein and Niels Bohr that the model one chooses to explain the properties of light depends on which activity the observer is interested in: for example, light traveling through a vacuum acts like a particle, but light passing through a grating acts like a wave. This counterintuitive and radical solution was not just a convenient compromise to explain a paradox of nature, however. Rather, as Bohr emphasized, this model described the inherent nature of light. He named the phenomenon *complementarity*, to emphasize the points that that *both* conceptualizations are necessary to explain light's various properties and that the dual model gives a fuller picture of its essential nature than either one alone.

Complementarity has profound implications for the concept of causality, at least as it operates in the submicroscopic world, for the claim that the behavior of light is caused or determined, in part, by whichever phenomenon the observing human is interested in examining would seem to contradict both our experience and Plato's belief that the essential properties of an object are static, as well as our belief that the basic constructs of nature are stable and unchanging unless altered by some external circumstance. Although this counterintuitive concept challenges and undermines our everyday experience, it remains accepted as accurate by the physics community as a necessary explanation of the observable properties of light.

Another counterintuitive idea that emerged at the same time was Einstein's famous equation, $E = mc^2$ (in which E is energy, m is mass, and c is the speed of light). Since the speed of light is a constant, that is, is a number that never changes, energy and mass are equivalent; once you know how much energy there is, you know the mass, and vice versa. However, this simple formula does more than just describe the interconvertability of energy and mass. It actually conflates or merges two concepts that everyday experience indicates are different into a single basic property of nature. Although this idea is conceptually different than the duality of light, it shows that even in everyday life we need two constructs, mass and energy, to explain some of nature's basic building blocks.

This idea that more than one viewpoint might be necessary to explain a situation best is reminiscent of Aristotle's belief that causality might sometimes best be constructed by using more than one level of his causal model.

The particle-like conception of light and energy reflects a dualistic, present-absent description of nature, that is, a categorical, dichotomous conceptualization. In contrast, the wavelike representations of energy and light follow probabilistic descriptions; the positions of these concentrations of energy are described in terms of likelihoods rather than absolutes. This offers a parallel to our description of cause as following categorical or probabilistic conceptualizations. Since both approaches have contributed to the fundamental understandings of modern physics, both approaches might also be understood as contributing to descriptions of causality as both categorical and probabilistic.

Another important implication of complementarity or duality for causal modeling is that indeterminacy does not equate to an inability to be specific. In fact, descriptions of light as a particle or wave, of the equivalency of energy and mass, and of the probabilistic nature of the quantum world have led to far-reaching advances in knowledge, and specific properties have been identified, described, and quantified for each of these scientific ideas. Duality does present challenges, but they are addressable. As with indeterminacy, knowledge of the limits of knowing each state actually increases the ability to describe accurately and, more importantly, verify the accuracy of the model's predictions. This can be seen as another contribution and strength of probabilistic reasoning.

Complexity is not without its problems, however, and the belief that simplicity is beauty is widespread, even among scientists. That some of the most accomplished physicists of the past seventy-five years should choose beauty ("simplicity is beauty," an aphorism sometimes referred to as the "law of parsimony" or as "Occam's Razor," after a fourteenth-century cleric) as a reason for believing that a "unified field theory" underlies the four forces of nature and is worth searching for speaks to the power of the human tendency to define simplicity as inherent in nature. The high value placed on simplicity and beauty by some scientists should temper those who describe narrative approaches (described in chapter 9) as arbitrary or inferior.

TIME, RELATIVITY THEORY, QUANTUM MECHANICS, AND CAUSALITY

Relativity theory and quantum mechanics developed simultaneously at the beginning of the twentieth century and have greatly influenced the research agenda in physics since then. In addition, they and derivative ideas such as instantaneous travel in space, black holes, and the Big Bang have captured the public imagination and influenced how many people perceive the world. In fact, these ideas have become so influential and are so powerful that any review of causality without their consideration would be incomplete.

While both quantum mechanics and relativity theory revolutionized twentieth-century physics, the public is much more aware of the latter and of its developer, Albert Einstein. Einstein's influence extended into politics, and his persona has shaped the public conception of genius. The construct of relativity theory has captured the public imagination, perhaps because of its relevance to such popular concepts as time travel.

The definition of time has engaged philosophers since the ancient Greeks. Until the twentieth century, most thinkers followed Aristotle and considered time to be an inherent, immutable quality of nature. An exception was Baruch Spinoza (1632–1677), who described time as a human construct that is imposed on the world by our experience and for our convenience.

Einstein's famous "thought experiment," which led to his description of the theory of special relativity in 1905, involved imagining what would happen if a person could ride on a beam of light. A person at rest behind the light rider would be "frozen in time" from the perspective of the light rider, since the light from him would not reach the person on the light beam. What would happen, then, if a person approached but did not reach the speed of light? Relativity theory predicted that a clock held by that person would register a different passage of time than a clock held by a person at rest. At everyday velocities, the difference in the passage of time between clocks traveling at different speeds is unobservable, but at fast velocities such differences would be significant, and measured time would differ. The revolutionary insight of relativity theory, then, is that *time is not a constant of nature* but depends on the velocity of the measurer and of the object being measured. Perhaps even more revolutionarily, relativity theory predicts that time does not have an inherent "forward" direction because it would go "backward" if a clock were able to go faster than the speed of light.

One reason that time's reversibility is counterintuitive is that it undermines the permanence of causality. It does this because the concept of causality requires that a cause precede its effect, and this sequence must be maintained over time if causality is to have permanency. If time is reversible, then an effect can precede its cause, much as a movie that is run backward makes it appear that causality is reversed. It is for this reason that the definition of cause proposed in chapter 1 posited that time is unidirectional. Is time's reversibility merely a thought experiment, or is it actually possible? Thus far, no experimental evidence has conclusively demonstrated that time actually can go in reverse, but many other predictions of relativity theory have been borne out by experiment. For the present, however, we must assume that time is unidirectional for a coherent model of causality to be constructed, at least given our current understanding of the concept.

However, the assumption and common-sense experience that changes occurring over a distance must take some time to occur are contradicted by experiments of quantum mechanical phenomena in which a change induced in one object *simultaneously* induces a change in another object some distance from the first object. That is, a change induced in one object induces a reciprocal change in another object *at the same instant*. This violates and has extraordinary implications for the concept of causality. If no time elapses between the two events, then our requirement that a cause must precede its effect is disproven. This is reminiscent of Hume's objection that in inductive reasoning two events can appear to be linked without any definitive method for demonstrating that such a connection has taken place. As with many of the ideas of quantum mechanics and relativity, it is not clear that these phenomena have counterparts in the macroscopic world of everyday experience, but the fact that simultaneity can occur at a distance in experimental situations presents a major challenge to the sequential notion of cause, one that cannot not be ignored. At present, though, we must posit certain relationships (such as time's unidirectionality) and build a model of causality from those pre-positions without a lack of certainty.

This is one of several places in this book in which competing, equally plausible ideas exist or for which the existing data support several models that appear contradictory. The issue is an important one and deserves a brief diversion and explication. What should be done when several models are plausible? In some situations, tests or experiments can be devised that refute some of the alternatives and eliminate them from further consideration. However, when more than one model appears to be accurate, three options are available. First, one can abandon the discussion as fruitless. Second, one can accept that contradictions exist and focus on those areas in which contradictions do not exist or are minimal. A third approach, the one adopted here, is to make the contradictions explicit, to identify and make clear the assumptions needed to continue the discussion, and then proceed. This approach accepts the notion that neither a coherent, single path of reasoning nor absolute knowledge is necessary before advances in knowledge can be achieved.

It is possible, though, that causal laws differ at the quantum and macroscopic levels, even though this would contradict our search for universals, simplicity, and beauty. As discussed in chapter 5, the constructs of nonlinearity, singularities, and power-law relations demonstrate that uniformity across all situations may not be a necessary characteristic of the natural world. The acceptance of the interconvertability of mass and energy and of the wave/particle duality of light suggests that the existence of simultaneity (an absence of time) at the quantum level and of the speed of light as a constant (which requires implying that time exists although it can be relative to the movement of the observer) in Einsteinian relativity theory may provide causal explanations for phenomena not now understood, even though the contradiction seems ludicrous to some. Engineers do not ignore the laws of physics, but they must apply them in ways that allow them to accomplish their task. Twentieth-century physics teaches us that we need not be stymied or paralyzed by the contradictions that seem to arise in the study of a difficult topic. The acceptance of complementarity and duality led to extraordinary advances in the understanding of both the basic building blocks and the macroscopic structure of the universe. Hopefully, the identification of contradictions in the study of causality will not stymie us but rather point the way toward developing models in which both sets of ideas can be accommodated.

RELATIVITY, PERSPECTIVE, AND COMPLEX INTERACTIONISM

The acceptance during the first third of the twentieth century of relativity theory, the wave/particle duality of light, Heisenberg's uncertainty principle, and Gödel's incompleteness theorem and the recognition of the inability to describe mathematically the forces that occur when three bodies interact illustrate that absolute knowledge is unattainable in many fields. Indeed, I will argue in chapter 10 that absolute causal knowledge is possible only in those disciplines that are "ecclesiastic," that is, those fields in which truth is given and received. There are differences in how causal knowledge is obtained in the empirical fields of science and the empathic fields of narrative knowledge such as history, politics, and philosophy, but this need not impede the search for causal knowledge in these fields. As discussed in chapter 3, very accurate predictions can be made in spite of the uncertainties inherent in the scientific laws that underlie the predictive models. The development of probabilistic reasoning and the field of probability statistics has enabled the empirical sciences to develop numerical descriptions that *estimate* the likelihood of a causal relationship.

However, we are still left with the question of whether these likelihoods can be validated or refuted. One way to address this important question and to develop criteria for causality is to examine instances in which consensus has developed (or failed to develop) in a discipline's search for causal mechanisms. In the remainder of this chapter, we will review how knowledge in geology accumulated to a point in which there is almost universal acceptance of a causal model. Chapter 7 explores examples in the biological sciences, and chapter 8 presents several examples in the epidemiological sciences. Later chapters will address similar questions in the narrative/historical disciplines (chapter 9) and religion (chapter 10).

GEOLOGY, CONTINENTAL DRIFT, AND THE PLATE TECTONIC THEORY

> Plate tectonics is a global theory—the first global theory ever to be generally accepted in the entire history of earth science.
>
> —Naomi Oreskes

Geology is the study of the structure and composition of celestial bodies and of the mechanisms by which they come to be. It seeks to explain phenomena that have intrigued humankind for thousands of years, such as the formation of the continents and the causes of catastrophic phenomena such as earthquakes and volcanic eruptions.

During the 1960s, a new theory, plate tectonic theory, became widely accepted as the causal mechanism explaining the formation on Earth of large, seemingly permanent phenomena such as continents and mountains as well as the cause of transient but powerful events such as earthquakes and tidal waves (tsunamis). Examining how this theory came to be accepted over such a short period of time by most geologists provides us with an opportunity to examine some of the principles of establishing causality.

Modern geology traces its beginnings to Charles Lyell (1797–1875), who proposed that the geological processes of the past can best be studied by examining the geological processes of the present, a theory called uniformitarianism. In 1912, Alfred Wegener (1880–1930), who had trained as a meteorologist, sought to explain why fossils of tropical flora and fauna were being found in geographic regions of the Earth that now have severe winters and why the rock tailings of glaciers can be found in areas that are now tropical. Wegener also noted, as had several cartographers in prior centuries, that the coastlines of several continents were mirror images that could be pieced together like a puzzle. For example, the Atlantic bulge of South America can be fitted like a puzzle piece into the Atlantic coast of Africa, and Great Britain can be fitted into the coast of France. His explanation for these observations was the *continental drift* hypothesis, which stated that the Earth was originally a single land mass (which he called Pangaea) that later broke into huge pieces (the continents) and "floated" apart. How this might have happened was not apparent to Wegener, and he died in Greenland in 1930 seeking evidence for the theory. Although this hypothesis was mentioned occasionally over the next fifty years, it was generally rejected by scholarly geologists, especially those in America.

In the mid-1960s, however, Wegener's basic idea became codified into plate tectonic theory and was widely and quickly accepted by geologists and other earth scientists as the explanation for the appearance of the Earth's surface. The theory states that the Earth's surface is made up of massive plates, each of which is eighty to 120 miles thick, and that these plates "float" on top of the molten magma that makes up the earth's mantle and core. Because this molten magma is heated by radioactivity and under great pressure, it continually forces its way to the surface of the Earth's crust through narrow, long rifts, or faults, in the Earth's surface, which are located in the middle of the great oceans. The hardening of the magma when it contacts the cool seabed results in the continual formation of several inches of new seabed per year on both edges of the fault through which the magma flows. As this new seabed is formed, it pushes the already existing plates that make up the seabed away from the rift. This pushes that plate into the plate that abuts it on the edge away from the rift, and this, in turn, transmits the pressure into the next set of plates, which begin at the continental coasts.

Because the continental plates are much more massive than the plates under the oceans, the seabed plates are pushed under the continental plates. The edge that goes under the continental plate then gets heated by the high temperatures deep in the earth and again becomes magma. The constant pressure on the continental plates also pushes the multiple plates that make up the continents into one another; if these collisions force one plate over the other, mountains form. If one plate moves suddenly in relation to the adjacent plate, an earthquake results. If molten magma suddenly moves through or between plates, volcanic eruptions occur. Thus, plate tectonic theory provides the causal explanation for several of Earth's large-scale geological processes.

How did this theory become so widely and rather suddenly accepted as the causal mechanism for these various geologic phenomena? In *Plate Tectonics*, Naomi Oreskes, the editor of the volume, enlists many of the scientists involved in the theory's development to tell the story from their individual points of view. She and they make the strong case that a wide range of data collected by a number of scientists in the early 1960s were put together by several individuals into a model that explained so many previously unexplainable phenomena that the theory's acceptance became a *fait accompli*. The data they accumulated included:

1. Seismographic evidence showing that earthquakes cluster in certain areas, many of which are along the coastlines of the continents
2. The discovery of hot midsea vents in the Atlantic, Pacific, and Indian Oceans, from which lava seeps
3. The discovery that the seabed is magnetized in strips that alternate every thirty-five kilometers
4. Evidence that the Earth's magnetic field reverses suddenly about every ten thousand years
5. Experimental models showing that solid plates moving on the surface of a sphere with a liquid core mimicked measurements made at many points on the Earth's surface with magnetometers and seismographs
6. The demonstration that islands closer to the midocean rifts through which magma was flowing were younger than islands farther away from those rifts

Item 3 was a crucial finding. It led to the hypothesis that the new sea floor is continually forming at midsea vents and that it is permanently magnetized by the Earth's prevailing magnetic field as it hardens. As such, it is a permanent record of the direction of the Earth's magnetic field at the time that part of the ocean floor was formed. When the Earth's magnetic field reverses, as it does every ten thousand years or so, the lava that hardens after the reversal is magnetized in the opposite direction to the part that formed prior to the switch. The discovery that the seabed is made up of bands of approximately similar width that alternate in the directions of their magnetization led to the conclusion that the seabed is forming at a relatively constant rate. This idea was quickly combined with the other observations listed above and with Wegener's observation that the continents could be fitted together into a single landmass. This led to explanations for a number of phenomena, including the formation of mountains, earthquakes, and volcanic eruptions. In addition to explaining plausibly so many phenomena of the Earth's geology, the theory led to predictions in fields as disparate as physics, paleontology, and marine biology. Many of these were quickly tested, and the positive results of these experiments and observations further confirmed the theory. In the face of such an explanatory mechanism that plausibly covered so many different phenomena and the confirmatory experimental and observational evidence, objections to Wegener's hypothesis quickly dissolved, and the theory was embraced by most geologists within several years.

In the example of plate tectonic theory, then, a number of elements and data sources convinced experts from disparate fields of the accuracy of this causal theory:

1. A large number of facts accumulated that could be linked in a plausible fashion (multiple lines of evidence; plausibility)
2. This model could be tested "experimentally" by seeking other confirmatory facts and doing experiments (confirmatory prediction)
3. Experiments and facts could be proposed that would disconfirm the theory (falsifiability)
4. The theory made new predictions that could be tested by observation and study, that is, experiments (further prediction and falsifiability)
5. Many natural phenomena and multiple disparate pieces of information could be explained by the theory (comprehensiveness)

The combination of multiple, unique supportive observations, confirmatory predictions, survival of falsification (the ruling out of alternatives), plausibility, and comprehensiveness gave and still give great explanatory power to the model. That many of the most respected leaders of the field quickly embraced the theory likely helped with its quick and universal acceptance as well. Findings in the past forty years have refined the understanding of mechanism, but the basic model has held up to scrutiny.

However, this is only one example of how a new explanatory scientific model came to be widely accepted. Many other examples could be cited that followed very different time courses, engendered greater or lesser resistance and disagreement, or never became universally accepted. Thus, plate tectonic theory does not provide *the* template for how consensus about causal mechanism develops in science—in fact, I believe it describes a path that is more the exception than the rule—but it does provide a clear-cut example of the ability to develop causal models using scientific methods.

REDUX

This brief review of several conceptual ideas that emerged in twentieth-century physics and mathematics and of the development of a specific causal theory in geology suggests that no single description or theory can describe how new causal knowledge develops or is accepted in the sciences. In part, this results from the subject matter of the specific branch of science. Some disciplines and questions lend themselves to experimentation and to the framing of questions and hypotheses that can be addressed by specific experiments. Others depend more on the accumulation of many observations.

Nevertheless, the construct of causality and the accepted view of causality in all scientific disciplines is strongly influenced by the strength of evidence supporting it and by the weakness of evidence not supporting or disproving other potential causal explanations. While other issues, such as the prevailing views of the larger culture, the ability of the explanatory model's proponents and opponents to communicate their views to others, the prestige of the individuals involved, and the strength of any already accepted model, influence whether and how quickly a particular causal model is accepted, the empirical approach of science rests on the belief that there is an actual truth and that verifiable facts (data) will ultimately convince people of the accuracy of any proposed approximation of it. Observation and experiment are the primary methods by which new knowledge is gained in the scientific disciplines, and scientists share the belief that knowledge accumulation over time is progressing toward a more accurate description and understanding of nature. This is a powerful approach, but it has significant limits, some of which have been enumerated thus far. Other limitations and strengths of the empirical approach will become clearer in chapters 7 and 8 as we review the establishment of causal knowledge in other scientific fields.

Causal modeling in the physical sciences rests upon the use of empirically derived data to explain phenomena using categorical, dimensional, or emergent logic. Different levels of analysis may require different models, and a coherent understanding of cause frequently requires analysis at multiple levels (predisposing, precipitating, or programmatic). For example, plate tectonics explains a categorical feature, earthquakes, as the sudden movement of one tectonic plate in relation to another (often sideways). These sudden movements result from the gradual increase in pressure put on the junction of these plates (a predisposing aspect, since other outcomes besides earthquakes are possible), and this in turn results from the molten nature of the core of the Earth, which is maintained by nuclear mechanisms. The sudden movement of one plate in relationship to another that results in an earthquake is nonlinear and appears emergent.

Thus, the molten nature of the Earth's core, the platelike structure of the planet's surface, and the distance of the Earth from the sun are all elements in the causal pathway for earthquakes, mountains, and continents. Multiple methods of reasoning and multiple levels of analysis contribute to the most comprehensive explanation for these phenomena. The unmistakable gain in knowledge over time in the sciences and the variety of methods used to accumulate this knowledge support the claim that no single model of causality can be put forth as universal. As will be seen in the next few chapters, other disciplines rely on different mixtures of these approaches and emphasize different methods of reasoning, but they too require a mixture of methods and models of reasoning.

7

EMPIRICAL: THE BIOLOGICAL SCIENCES

Diversity is maintained because of the global uncertainty of local uncertainty.

—Simon Levin

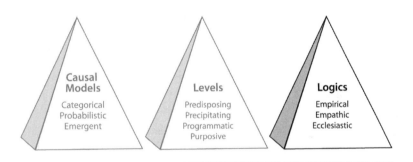

The biological sciences offer the opportunity to examine the application of the empirical method (facet 3) to the study of causality because many of the issues explored by biologists are amenable to experiment. This chapter will use three topics, the nature-nurture debate, the application of causal knowledge to the goal of eliminating specific infectious diseases, and the emerging discipline of ecology, to illustrate the application of the empirical method and its interaction with facet 2's level of analysis and facet 1's causative models. The chapter will end with brief discussions of two topics that have been discussed before: the directionality of time and the concept of chance.

GENE-ENVIRONMENT INTERACTIONS

The nature-nurture and gene-environment controversies can be boiled down to a simple question: how much of any specific characteristic of an organism is generated by innate mechanisms, and how much is induced by the environment? At first blush, some examples seem to have straightforward answers. For example, height in humans is among the most heritable of features and might, therefore, be thought of as predominantly "nature" in origin. Choice of occupation, on the other hand, would seem to be predominantly "nurture" influenced, since many occupations of today did not exist one hundred years ago and many occupations of yesteryear do not exist today.

However, even these examples reveal the challenge of the nature-nurture question. Studies find that about 40 percent of occupations can be linked to shared genetic factors. This suggests that the skills and personality factors that influence job choice and performance are shaped by genetic endowment. Individuals with above-average mechanical skills, for instance, would have been more likely to become metalsmiths in the past and are more likely to become plumbers, electricians, or computer technicians today than individuals with lower than average innate mechanical skills. Likewise, individuals with above-average mathematical skills are more likely to become bank tellers, accountants, or computer programmers today. Thus, the answer to the nature-nurture question is usually not "only nature" or "only nurture" but rather a description of how much each contributes causally to the specific issue.

The interaction between nature and nurture in determining the height of adults is even more illustrative and became evident to me several years ago during a trip to Japan. I am five-foot-four, but I never much think about the height of others. However, while riding in an elevator at a scientific meeting in Japan one afternoon, I noticed that I was peering down on the heads of most of the other adults in the elevator. This was an unusual enough experience that I immediately began wondering whether there was anything unique about my co-riders. The only thing I could figure out was that all of the individuals in the elevator were either my age or older. Later, during another elevator ride, I noticed that the Japanese attendees who were younger than I were either taller than I or at face level, as is the case in the United States. The difference in the two experiences seemed to be the age of my fellow riders, but this was puzzling since I had learned previously that height is one of the most genetically determined of traits. How could this be?

It is universally accepted that the heritability of height is greater than 0.9, placing it among the most genetic of traits. Supporting this is the finding that the heritability of height in children adopted at birth correlates strongly with the height of their birth parents, not their adoptive ones. However, the distribution of height in Japan differs between cohorts born after World War II and before it, even though height is distributed in the Gaussian bell-shaped curve discussed in chapter 4 in each cohort. How can we explain the dramatic height difference between these two generations? From other research we know that mean adult height has increased gradually over the twentieth century, which is attributable presumably to improved nutrition. Thus, the likely explanation for the different final heights in the two Japanese cohorts is that an environmental event, for example, a change in diet (most likely the availability of more calories and protein), strongly influences final achieved height. That is, the potential height of an individual can only be fully reached when the environment provides adequate nutrition (and perhaps other factors) for the genetic endowment to reach its full potential, that is, to be fully expressed. If all individuals are exposed to the same environment, then the "place" a person occupies in the distribution of height in the population is genetically determined. If the amount of food or other necessary elements that individuals are exposed to during development is less than maximal or varies significantly, then the environment plays a significant role.

This discussion has a serious flaw and raises a difficult question: does this argument *prove* that improved nutrition is the cause of the observed rise in height in Japan during the last sixty years? Mean height has increased during this period, and there is good evidence that food supplies were limited during World War II. But how do we assess whether these are linked causally and not just coincidentally? Bradford-Hill's criteria, listed in chapter 1 and discussed in more detail in chapter 8, are a help. Examples of a similar association in other countries or geographic areas would be supportive of a causal role, but the possibility that some other environmental cause or causes was responsible would remain. How about Popper's criterion of falsifiability? It is too high a standard because the change in height is a historical event that cannot be repeated, and it would be unethical to perform an experiment in humans in which the calorie or vitamin/mineral content of food is restricted in one group of infants and not another. Furthermore, even if one examined data from "naturalistic experiments" in which low calorie intake or nutritional deficiency occurred in several geographic areas, it is still very possible that some other environmental characteristic associated with the nutritional intake difference (stress, for example) is the actual causative agent. Studies in animals could generate data that would either verify or reject (falsify) the hypothesized linkage, but nutritionally deprived animals live longer than animals given liberal feedings of laboratory chow, so some other mechanism might be operative. We thus have several lines of supporting evidence—the finding of similar associations in several different geographic locations, ethnic groups, and periods in history; the existence of biological mechanisms that link the availability of nutritional requirements to periods of developmental vulnerability in humans and other animals—but must rely on the inductive causal reasoning that Hume criticized. If absolute proof of causality is not possible in this rather simple example, then proof of causality in more complex biological situations, especially those that cannot be manipulated experimentally, will be even more challenging.

Since genetics is the primary determinant of final height if adequate nutrition is available, genes are best considered a predisposing cause; the environment acts as a precipitating cause. When maximal nutrition is available, height appears to be primarily genetic, as indicated by its hereditability of 0.9. In this instance, the precipitating cause is able to be fully operative. However, when the environment, that is, nutritional availability (this includes nonfood issues such as sunlight exposure) is not maximal, the predisposing causal role of the environment becomes clear. However, the issue is even more complex, because growth occurs during a specific period of development, and other environmental events such as war, drought, or toxin exposure play a role. A programmatic level of analysis provides an integration of these many factors and, in some circumstances, provides the most comprehensive approach.

This is a good illustration of the erroneous conceptualization of "nature" and "nurture" as opposing, fully independent, mutually exclusive, determinative causal agents. Recognizing nature and nurture as interacting factors that mutually and independently contribute to attained height in a complex way is a more complex formulation, but it provides more explanatory power and best accounts for what is observed in nature. The same can be said for understanding the distribution of blood pressure, personality traits, intelligence, and fasting blood sugar in the population—the evidence is strong that both genetic endowment and environmental influences contribute significantly to the organism's constitutional makeup in relationship to these characteristics. The either/or construct of the dichotomous causal model is clearly not applicable to these aspects of biology; whether the linear relationship of the dimensional model or the nonlinear relationships of the emergent model best represents the relationship between the variables in any given process will vary, but in these examples multiple factors are operative, and nonlinear modeling is likely to capture best the causal relationships in each system.

Some biological phenomena do have simple explanations that follow categorical logic and have either a genetic or environmental etiology. For example, Huntington disease, a rare form of dementia, is caused by an abnormal gene on chromosome 4. It is inherited as an autosomal dominant trait; that is, people born with the abnormal gene will inevitably develop the disease if they live long enough (on average it begins around age forty, but rarely the disease begins in a person's sixties). This abnormal gene is the cause of the disease, and the disease occurs only in those with the abnormal form of the gene. It is an example of categorical genetic causality (even though the variation in age of onset is related to the length of the abnormal portion of the gene). However, most diseases have multiple, interacting causes. The remainder of the chapter will examine several examples of this more complex causality.

INFECTION AS A CAUSAL AGENT

Theoretically, infectious diseases should be straightforward, environmentally caused illnesses. In fact, though, most are best understood as resulting from complex gene-environment interactions. As such, they serve both as instructive examples of multilevel, causal relationships and provide further insight into the limitations of the dichotomous yes/no model advocated by Galileo that has served science so well for four hundred years.

Infectious diseases such as influenza, plague, and smallpox (before it was eliminated from the human population) cause time-limited, specific diseases. Although each is caused by an organism that is always present somewhere in the environment, the number of people who are ill from the viruses that cause smallpox and influenza or the bacteria that causes plague varies widely over time and between places. The time periods during which there is a rapid increase in the number of individuals affected are referred to as epidemics, and the epidemics caused by each organism follow different and unique patterns. How can this be understood from a causal perspective when all are caused by infectious agents?

In part, the variability in number of persons who are ill over time is explained by the fact that when individuals are infected, their immune system forms antibodies that eliminate that specific organism. As a result, if individuals are infected a second time their immune system mounts a quick response that usually eliminates the agent and prevents the person from becoming systemically ill a second time. This means that the number of people who can become sick from that organism at any one time varies: only those who lack immunity are vulnerable to becoming ill from that specific infectious agent.

A second issue contributing to the epidemic nature of some infectious diseases is the ease of spread or contagiousness of the organism. When an agent that is highly contagious is introduced to a group of individuals who have not previously been exposed to it, then most will become infected and fall ill. Even in epidemics, though, not everyone is exposed to that organism or becomes ill from it. These individuals remain vulnerable to infection if they are exposed to the agent in the future. Furthermore, some organisms also infect nonhuman animals and thus have "reservoirs" in which those organisms are maintained. When those infected nonhuman animals come into contact with susceptible humans, the disease-causing organisms spread from them to infect those humans who have not previously been infected.

A primary explanation for the epidemic nature of some infectious diseases, then, is that people who are vulnerable can become infected and ill but that after this occurs, the disease becomes rare until a new group of susceptible individuals is available. In between epidemics, the infection rate remains low, and the organism survives by infecting nonhuman species ("reservoirs"), by infecting the small number of people who are still vulnerable, or by inhabiting a small number of individuals (humans or nonhumans) who do not develop immunity to it but are either not ill or have minimal symptoms (so-called carriers). After a period of years, the number of people who have never been exposed to the organism, often children or young adults, increases until there is again a large, vulnerable population. When the organism finds its way into that group, an epidemic ensues.

While this general description describes the genesis of many infectious disease epidemics, each organism has unique features that shape the particulars of its life cycle and the disease or diseases that it causes. That is, the biology of each causative organism is quite different. Therefore, gaining an understanding of causal mechanisms associated with any particular organism and any particular infectious agent also requires knowledge of the details of that organism's genome and behavior, knowledge of the host's behavior (and genome), and knowledge of the environment.

For example, the plague bacillus, *Yersinia pestis*, is spread to humans via flea bites. The fleas become infected by biting infected rats. Thus the rat is the reservoir. For humans to develop plague, they must have close contact with rats because fleas cannot live away from a warm animal for very long. Therefore, an epidemic of plague in humans requires a large number of humans who have not previously been infected with plague, a large number of rats infected with the *Yersinia* bacillus, and environmental conditions that place humans and rats in close proximity (that is, environmental conditions that increase the likelihood that humans will be bitten by the fleas that are carried by the rats). While human-to-human spread of plague occurs, it is rare. This is more likely to occur when many people are infected, as during an epidemic, but it does not contribute significantly to the absolute number of people who become ill, that is, the epidemic nature of the illness.

Influenza is a respiratory illness caused by the influenza virus. One of the unique characteristics of this virus is that two of the genes that control its ability to infect cells in humans have a tendency to mutate. That is, the genetic makeup of this virus includes mechanisms that "encourage" rapid mutation, that is, make it more likely than by chance. As a result of these mutations, strains of influenza virus emerge that are genetically different enough (that is, "new") so that people who have had a prior influenza infection are not immune to the mutated virus. These mutations occur while the organism is maintaining itself in nonhuman animal reservoirs, thought to be certain fowl and pigs. The greater the number of chickens, ducks, or pigs, the higher the likelihood that a new strain will emerge. Thus, influenza epidemics arise because the virus is maintained in organisms that humans have domesticated, raise in large numbers, and, in some parts of the world, live in close contact with; because the biology of the virus includes an innate mechanism that makes mutation likely; and because the virus is easily spread by human contact, either through touch or aerosol (sneezing and coughing). It is this combination of the innate biology of the organisms, the innate biology of humans (vulnerability of respiratory tract cells to the infection), and environmental events (domestication of chickens and pigs, human crowding in the winter, the human custom in many cultures of touching each other by hand) that enables influenza epidemics to occur.

Other patterns of human behavior contribute to the likelihood that an epidemic will occur. Bringing together large numbers of people who have not previously had contact in places such as college dormitories, summer camps, and army barracks increases the likelihood that large numbers of never-exposed individuals will come into contact with pathogens such as the influenza and polio viruses for the first time, and this increases the likelihood of spread from one human to another. The widespread use of international travel further increases the likelihood that organisms will be carried to unexposed and therefore vulnerable groups. Another human behavior that can influence the likelihood of disease spread is the manner in which garbage is stored, since this can affect both the number of rats and their proximity to humans, which in turn increase or decrease the likelihood that fleas infected with the plague bacterium will have human contact. As already noted, the greater the number of fowl, the higher the likelihood that the influenza virus will mutate into an antigenically different virus, and the more contact that humans have with the fowl they raise, the greater the likelihood that a new virulent mutant virus will spread into the human population and infect a large number of people.

Furthermore, individual humans vary in their vulnerability to developing infectious (and other) diseases because of variations in the genetic makeup of their immune system. Three examples illustrate the point. Those who develop meningitis from the bacterium *Neisseria meningitidis* are more likely to have specific variants of a gene that encodes the mannose-binding protein and complement factor D than people who are exposed to the bacterium but do not develop meningitis. In fact, this bacterium is present in the nasal mucosa of many individuals, and the vast majority suffers no ill effects. Thus, the highly fatal disease neisseria meningitis requires both exposure to the bacterium and a genetic vulnerability that decreases the ability of the immune system to eliminate the organism effectively—it is the combination that should be considered causal. Similarly, many of those who survived the great Black Plague epidemics of past centuries are thought to have survived because they were able to muster a different immunologic response than those who died from the infection. A contemporary example is the long-term survival of a small percentage of those infected with the HIV virus before there were effective treatments; in some individuals this is explained by the presence of a genetic variation that limits how the virus spreads from cell to cell within that individual.

Thus, epidemics of plague, influenza, smallpox, and HIV/AIDS can be understood as resulting from biological ("nature") and environmental ("nurture") features of both the organism and the "host" human. Epidemiologists call this interaction the host/agent/environment triad to highlight the importance of each element of these interacting systems and to emphasize that each has a role in the causal pathway and outcome. The Aristotelian approach would classify genetic makeup (a vulnerability in the neisseria example and a protective factor in the HIV example) as a predisposing cause and the infectious agent as the precipitating cause. For a new variant of the influenza virus, the genetic mutation and the virus are precipitating causes; the human behaviors of raising fowl, touching other humans, and living in close contact are predisposing causes. For plague, the bacillus is precipitating, but the passage via rats and fleas, a necessity in the life cycle of the human infection, is more complex. Galileo would have called these necessary but not sufficient causes, but they have elements of both predisposing and precipitating cause; without close human proximity to large numbers of rats (and their fleas), an epidemic of plague would not happen—a feature of precipitating cause. However, rats alone, even in high numbers, are not directly causal and so can also be seen as predisposing.

The epidemic nature of each of these illnesses can also be understood causally at a programmatic level of analysis because it is the weaving together of the biology of the causal organism; specific environmental circumstances as varied as weather, human social custom, and the raising of certain animals for food; as well as the biology of the individual host and of the species that lead to outbreaks of these illnesses. Empirical data help identify both predisposing and precipitating relationships among individual causal elements and inform a weaving together of these individual understandings into a programmatic understanding of the causal web. Each level of analysis contributes to this understanding, and any single relationship, predisposing, precipitating, or programmatic, might lead to an intervention that can interrupt the causal chain.

GENE-ENVIRONMENT INTERACTIONS AND THE VULNERABILITY TO DISEASE

While the neisseria and HIV examples address relatively uncommon genetic predispositions and disorders, there are other examples in which host genetic factors influence more common disorders, examples more akin to the attainment of final height. For example, recent research has linked the gene abnormality that causes a relatively uncommon disorder, cystic fibrosis, to sinusitis, an infectious disease that is one of the most common of human afflictions.

Cystic fibrosis is caused by mutations in the gene that directs the formation of chloride channels in cell membranes. Dysfunction of these channels results in the impaired performance of many organ systems, most prominently the lungs (resulting in a difficulty in clearing mucus from the bronchial tree and thereby increasing the risk of lung infection) and the pancreas (resulting in impaired digestive function). The disease is often fatal in childhood or young adulthood because repeated bouts of pneumonia irreparably damage the lungs. The gene that is affected, the CFTR gene, is located on chromosome 14 and operates in an autosomal recessive fashion. That is, cystic fibrosis occurs only when the CFTR genes inherited from each parent carry the causal mutation. Thus, the disease state of cystic fibrosis requires the inheritance of two abnormal copies of the gene.

Intriguingly, it has recently been shown that having only one mutated CFTR gene increases the risk of chronic sinusitis. This single abnormal CFTR gene does not cause the sinusitis. Rather, it increases the likelihood the nasal sinuses will become chronically infected when a person is exposed to infectious agents that are ubiquitous in the environment. Thus, the precipitating causative agent of the sinusitis is environmental (common bacteria), but the predisposing cause, the risk factor that increases the likelihood of developing sinusitis, is genetic. Not all individuals with a single mutated CFTR gene suffer from chronic sinusitis, and most individuals with chronic sinusitis do not have a single mutated CFTR gene. Furthermore, the abnormal CFTR gene is more common in Caucasians than other groups, so it would more likely contribute to chronic sinusitis in whites than other ethnic groups. Clearly, the presence of this gene variant alters the body's ability to handle the infection and increases the risk that an infection in the sinuses will become chronic. Thus, the CFTR gene is a predisposing cause of sinusitis when one copy of the gene is abnormal, but it is a precipitating cause of the disease cystic fibrosis when a person inherits two abnormal copies of the gene, one from each parent.

Another example of how genetic variation can influence vulnerability takes us away from infectious disease but nicely illustrates the predisposing nature of both genetic makeup and environmental exposures. The SCN5A gene directs the production of a component of the sodium channels that are found in the cell membrane of each heart muscle cell. These channels open and close and by doing so provide a potential direct connection between the interior of the cell and the surrounding exterior environment. When they open, sodium quickly moves into the cell. This sudden movement of sodium ions into a heart muscle cell causes it to contract, and it is the coordinated contraction of many cells that results in the pumping action or "beating" of the heart.

A number of variations have been found in the SCN5A gene. In one, a single nucleotide is changed; this results in an amino acid change from serine to tyrosine. This change makes the sodium channel prone to opening more quickly than usual. If this happens, the usual coordinated firing of heart cells may be interrupted, resulting in an irregular heart rhythm. This irregular heart rhythm interferes with the efficiency of the heart's pumping action, and the resulting arrhythmia initiates a downward spiral in which the heart itself does not receive enough blood, its pumping action further deteriorates, blood circulation ceases, and death ensues. Thus, the change of a single amino acid (out of a billion) in one single gene (out of approximately 21,000) predisposes an individual to an arrhythmia that can be fatal.

The likelihood that sodium channels open "off-time" can also be increased by certain environmental events. For example, diuretic medications ("fluid pills"), by lowering the amount of potassium in the blood, and the medication quinidine, which directly affects the function of potassium channels, affect the speed at which the electrical charge is conducted through the heart and increase the likelihood that an arrhythmia will develop. Thus, individuals with variants of the SCN5A gene who take these medicines are at even greater risk of developing an arrhythmia.

However, some individuals taking these medications who do not have this genetic vulnerability suffer the arrhythmia (perhaps they have other predispositions, or the potassium is lowered enough to be precipitating), and not all individuals with the genetic vulnerability who are exposed to these drugs develop an arrhythmia. Thus, both the genetic vulnerability and the environmental exposure are predisposing in one sense. However, everyone will develop the fatal heart rhythm if the potassium blood level is lowered enough by medication, other illness (for example, diarrhea), or diet. Hence, the diminished availability of potassium can be the precipitating factor of a heart arrhythmia and death when the degree of potassium depletion reaches a threshold below which cells cannot function normally. Low potassium is thus both a predisposing and precipitating factor, and it in turn is predisposed to by a variety of factors, some environmental and some genetic.

In one study, the tyrosine to serine substitution in the SCN5A gene was found in 19.2 percent of persons of West African and Caribbean descent, 13.2 percent of African Americans, and none of 511 Caucasians and 578 Asians. In a single family studied in depth, those who inherited this changed gene (called a polymorphism rather than a mutation because it occurs in more than 1 percent of the population) were significantly more likely to have a specific change on their electrocardiogram indicating slowing of electrical conduction between the ventricles (prolongation of the QT interval), a known risk factor for developing an abnormal heart rhythm. However, as the authors of this study note: "most of these individuals will never have an arrhythmia because the effect . . . is subtle." This genetic variant (not traditionally thought of as an abnormality) is thus another example of predisposing genetic causality: those with this variant are at increased risk of developing an arrhythmia because of it, but most of those with the variant do not suffer any adverse outcome. Taking drugs such as a diuretic or quinidine may precipitate an arrhythmia in these vulnerable individuals, but not always.

These examples of gene-environment interaction illustrate the application of the predisposition/precipitation distinction. They also demonstrate that predisposing causes can follow either categorical (the SCN5A gene polymorphism is either present or not) or dimensional (in all individuals, the lower the blood level of potassium, the greater the risk of an arrhythmia) logic. Precipitating causes can likewise be either categorical or dimensional. Cystic fibrosis is caused (in a precipitating fashion) by the presence of an abnormal chloride channel (categorical) that results from having two copies of the abnormal CFTR gene. Everyone who has two copies of the abnormal gene will have the disease (a categorical cause). The use of quinidine can induce an arrhythmia (a precipitating cause) in a person who is predisposed by having a copy of the SCN5A gene that contains the predisposing polymorphism but also in others with no known predisposition.

These examples also demonstrate that distinguishing between different levels of analysis (facet 2: predisposing, precipitating, programmatic, purposive) and differing models of causal logic (facet 1: categorical, dimensional, nonlinear) can help improve communication about and our understanding of the complex web of causality. The predisposing factors discussed in the above examples are clearly a part of the causal chain that leads to sickness or death, but many individuals with these factors will never have a problem. This can lead to skepticism, rejection of the causal linkage, and failure of the application of this knowledge, and this can be and has been a significant source of confusion to the public, the policy maker, and even the scientist. For example, the identification of risk or vulnerability factors has played an important role in preventing disease. Some prevention strategies such as vaccination and water fluoridation have targeted the whole population because most individuals are at risk for measles and dental caries. On the other hand, risk factors such as high cholesterol, high LDL cholesterol, and high blood pressure increase the likelihood that an individual will have a stroke or heart attack. Dramatic declines in heart attack and stroke rates have followed interventions that address these risk factors in individuals with these predispositions, a further support for the claim that the causal chain has been altered (à la Koch's postulates). However, lowering cholesterol by itself does not always lead to these benefits: recent studies of newer cholesterol-lowering agents have found no diminution in death, heart attack, or stroke rates. Recognizing that cholesterol and high LDL are predisposing rather than precipitating causal factors and therefore other biological mediators must be important in the causal chain helps explain this surprising finding.

The development of techniques to examine rapidly and inexpensively the genomes of large numbers of individuals is already identifying many potential genetic and environmental elements in the causal chain of disease and normal biology. As the examples discussed in this chapter illustrate, such discoveries are likely to be a source of conceptual confusion because many people equate the concept of cause with the categorical logic of "yes/no." It is my hope that the use of terms such as "predisposing cause" and "precipitating cause" will clarify discussions about causality among both experts and the public at large and improve acceptance of potentially beneficial interventions.

For scientists, these distinctions in the level of analysis and logic are important because different mathematical and statistical methods are used when describing and explaining causal relationships that follow these different models. For policy makers, politicians, medical educators, and public health advocates, recognizing that the concepts of likelihood and vulnerability, which underlie predisposing causes and dimensional logic, are difficult for many to understand might lead to education programs than can improve understanding and acceptance of interventions that address predispositions and provocations. This might be especially useful and important in cultures in which one mode of logic (categorical or dimensional) predominates.

Making such distinctions might also help explain why some interventions must simultaneously be directed toward different levels of causal analysis (facet 2) and utilize different models of causal logic (facet 1). For example, the incidence of malaria can be lowered and therefore cases prevented by the use of netting when sleeping in areas in which the disease is endemic. This is because it decreases exposure to the vector (mosquitoes) and therefore the parasite; in other words, it addresses the predisposing cause (mosquito bites) and the precipitating cause (the parasite). However, once a person is infected, direct treatment of the precipitating cause with antiparasite medication is necessary. Being clear about these different levels of analysis and different models of causal logic might help those who design interventions, fund them, implement them, and receive them understand why multiple approaches are necessary if the goal is to decrease the morbidity and mortality caused by malaria. Emphasizing one to the exclusion of the other is likely to lead to the failure of the overall goal of malaria prevention. Conceptual clarity might increase the likelihood of success by helping planners tailor when and where one approach needs to be emphasized more than the other.

Using these conceptualizations might also temper unrealistic expectations and emphasize the difficulties inherent in devising a strategy that addresses multiple levels of cause. Lowering the level of a risk factor (high blood pressure or high LDL cholesterol, for example) does not eliminate the occurrence of heart attack and stroke. Some individuals who never smoke cigarettes develop lung cancer. That is, intervening at the level of the public does not necessarily benefit all individuals. From the point of view of an individual who develops the undesirable outcome (for example, stroke), the intervention has been a failure, even though it might be a public health success because incidence rates in the population at large are decreased. Clarity over these issues has the potential to help explain these complexities and both maximize success and lessen disappointment. More accurate labeling will not solve the challenges of explaining complex ideas, of course, but can both highlight the challenges faced when trying to interrupt the causal chain and improve the quality and accuracy of public and professional discourse.

The discussion thus far has touched on programmatic causality but has highlighted the predisposing and precipitating levels of analysis of facet 2. The remainder of this chapter will reverse this emphasis. The purposive level of causality, referred to as teleology when it is applied to inanimate objects, will be highlighted in part to demonstrate that this level of analysis is not restricted to narrative or ecclesiastic fields.

ECOLOGY

> The subject of population ecology can be very complicated. But, as we do in any science, we begin by assuming that it is simple.
>
> —Vandermeer and Goldberg

Ecology is the discipline that seeks to integrate what is known about the animate and inanimate worlds into explanatory models of the environment and to identify causal mechanisms that bridge multiple complex systems. Its goal is no less than a coherent theory of how nature operates on the planet Earth. Although this issue has interested scientists for centuries, the discipline itself is relatively new. It thus provides an interesting case example of how knowledge about programmatic causality develops.

Ecologists have taken multiple approaches in their efforts to understand the complex relationships between the living (biotic) and nonliving (abiotic) elements of the environments they study. As the quote at the beginning of this section suggests, one strategy has been to study small, well-circumscribed systems; identify relationships that exist among their flora, fauna, geography, and weather; and then determine if similar relationships exist in larger geographic areas. One such finding from the study of island ecosystems is that the number of species increases as an island gets bigger and that this relationship can be described mathematically as a power or exponent (ranging between 0.1 and 0.4) of the size (area) of the island. For example, if you compare two islands that differ in size by a factor of ten, you would expect to find between 1.58 (that is, $10^{.2}$) and 2.51 (that is, $10^{.4}$) times more species on the larger island. The regularity of this relationship across many islands suggests an underlying causal mechanism and has led ecologists to ask whether the same mechanism is active in larger, nonisland ecosystems.

One obvious reason for focusing on smaller isolated geographic settings is that relationships are more easily identified when there fewer species to study and less variation in geography and weather. By limiting the number of contributing elements and thus complexity, this approach has the strength of simplifying the search for general principles, but it runs the risk of oversimplifying the complexity inherent in large systems and therefore underestimating or missing aspects of the causal ecological web. Thus, "assuming that it is simple" has its advantages but also carries the risk of limited applicability to significantly larger systems.

Studying single islands and coming to some understanding of their ecology also leads to the ability to consider each as a natural "unit" or system and provides a basis for comparing one system to another. This is a next step in the identification of relationships and broad-scale "forces" that act at multiple sites, and it could help identify underlying causal forces acting at the level of the macro environment. This is the approach Darwin utilized when he identified similarities and differences across the Galapagos islands and integrated this information with his knowledge of other biological systems to develop and support his theory of natural selection. More recently, this approach has been taken by Robert McArthur and E. O. Wilson in their book *The Theory of Island Biogeography*, in which they study how the makeup of island biota is influenced by size of the island and its distance from the mainland and then use this data to generalize about the impact that newly introduced species have on the distribution of existing species. They interpret the results as supporting the broad concept of biota in equilibrium and use this idea to derive generalizations about how biotic systems respond to change. From these generalizations they conclude that mechanisms for maintaining equilibrium operate at the level of the system at large. This sets the stage for the identification of these mechanisms.

The strategy of using knowledge gained from the study of multiple, individual interactions to make generalizations about larger systems has a long history of success in biology and other sciences, but it has been difficult to transfer to the identification of causal mechanisms at the programmatic level. One reason may be that the relationships among individual variables are often dichotomous or linear, while those that exist in complex systems tend to be nonlinear. The development of tools to test potential causal mechanisms such as equilibrium maintenance is a goal yet to be realized, but it will be necessary as long as empirical support for proposed causal mechanisms remains the standard of scientific reasoning.

Similar potential benefits and cautions apply if one begins by focusing on molecular-level relationships of an ecosystem and using this knowledge as a starting point for identifying generalizations about multilevel and multispecies relationships. A nice example of this approach is a review of soil system dynamics by Young and Crawford. They explain the aggregation of clay soil at the atomic level in terms of electrostatic and van der Waal forces, at the cellular level by the presence of gluelike substances secreted by bacteria and fungi and by the physical contribution of these rod-shaped bacteria and fungi, and at the macroscopic level of multicellular biota by the binding together of soil particles by plant roots. Intriguingly, soil particle aggregation follows a power-law distribution in the range that fungi and bacteria are found to contribute to aggregation.

This model can be expanded by describing the relationships among belowground (soil) and aboveground elements of ecosystems that support and differentiate lush and sparse environments. Factors to consider include physical characteristics of the soil, mineral availability in the soil, and the contributions of plants and animals. This linking together of the abiotic physical makeup of soils (that is, physics), the biology of the microbes that inhabit the soil (microbiology), the biology (botany) of the plants, and the behavior (zoology) of animals demonstrates how the merging of information about smaller system elements can begin to capture the complexity that exists at the level of the broader system. Such a systemwide understanding requires the integration of information and expertise from a variety of disciplines that are now considered distinct and a combination of methods and approaches that is not generally in the repertoire of a single or even a small handful of individuals. Whether causal explanations can be derived and validated from multilevel descriptions such as this example remains to be seen, but this is the path that science has taken in the past. Chapter 8 will examine methods proposed to overcome this problem, but it is worth pointing out here that a demonstration that multiple systems follow the same pattern (convergent validity), the ability to describe relationships over larger and larger canvases (coherence), and the ability to demonstrate that predicted relationships exist in locations or systems not previously studied (predictive validity) are among the empirical methods that are available. The studies of islands in equilibrium and disequilibrium by McArthur and Wilson and by Darwin in the Galapagos are examples of using naturally occurring phenomena to generate and later verify or validate broader causal mechanisms.

However, as network theory has shown, the larger that systems become, the more likely they are to have unique elements. This makes it more difficult to validate system-level causal hypotheses and further illustrates the challenge introduced by the increasing number of nonlinear relationships that exist as system size and complexity increase. This again raises the possibility that the identification of causal mechanisms operating at the system level will require analytic strategies that are unique or different than those used to study the relationship between two or several variables or between small and large ecosystems.

Δ Δ Δ

Starting with small, more basic elements of a system and incorporating other information into a broader understanding is sometimes
referred to as the "bottom-up" approach. It seeks to understand the
whole by first understanding its parts. In the "top-down" approach,
the initial focus is on the system as a whole and the identification
of relationships that exist among its constituent parts, that is, moving from the direction of understanding the whole to understanding how its parts relate. The studies of islands and the identification of replicated relationships between species number and area
is an example of a top-down description; the movement from an
electrostatic understanding of soil dynamics to a broad understanding of the role of different species is an example of the bottom-up
approach. Each has strengths, and it is likely that broad understandings of causal mechanisms, especially programmatic ones, will require
an accretion of knowledge from both approaches. Ecology seems to
be at the beginning of this process of identifying causal mechanisms
that explain how large systems came into being and how they continue to operate. It is certainly too early to conclude that the issues
are so complex that causal mechanisms either cannot be identified
or do not exist.

One example of the top-down approach to ecology that begins
at a much higher level of abstraction can be found in Simon Levin's
book *Fragile Dominion: Complexity and the Commons*. After reviewing a
number of research themes in ecology, Levin abstracts six general
patterns and interactions in ecological systems:

1. Ecosystems consist of regular patterns of interaction and response
   that human beings can codify, recognize, and study.
2. These macroscopic patterns usually develop over long periods of
   time and build upon changes that occur at the local level.
3. Historical accidents play a significant role in determining patterns of development and stability.
4. Chance, operating through accident, embodied developmental
   mechanisms, environmental events, and stochastic processes,
   provides the substrate for change.

5. The interaction of many biological systems results in an ecosystem that encourages diversity.
6. Complex adaptive systems rely on diversity, nonlinearity, and hierarchical organization.

These principles exemplify the types of generalizations that can only be identified when a top-down approach is taken. One clear theme is that nonlinear, emergent relationships play an important role in ecological systems. The role of chance operating as a predisposing cause is another. Levin notes that some alterations that occur in a single area of a system can become common locally because they are neutral, that is, provide no advantage or disadvantage, but later, when other changes occur, become widespread because they offer an advantage. In this instance, the environment is operating as a provoking cause and natural selection as a programmatic cause. The widespread dissemination of such a change would appear to be nonlinear because the predisposing change would not have been observable even though it had gradually accumulated, and the emergence of the adaptation would seem to occur over a short period of time, suggesting a sudden, categorical, nonlinear cause because the adaptation spreads quickly when the second circumstance occurs. In chapter 11, this is discussed in relationship to Eldridge and Gould's "punctuated equilibrium" theory of evolution.

This model of change, in which an initial minor and often unobserved event or change precedes and is necessary for a later large-scale change, can be difficult to discover and is not what people usually consider when seeking to establish the causes of a change in the environment. A careful analysis of the details of the specific situation may be required to reveal the necessary role of the predisposing (and seemingly "minor") change, while a broad perspective is often needed to observe the large-scale change. However, this does not mean that the direction of causality is from large to small. In fact, data from ecological systems reviewed by Levin demonstrate that the direction is often from small to large.

This insight is also demonstrated by a generalization that can be drawn from the earlier discussion of gene-environment interactions. Environmental extremes tend to exaggerate the influences of both genetic variability and environmental selection. If environmental extremes persist for several generations, those genetic traits that are more adaptive in the extreme environment will be passed on to off-spring because they will be more likely to reproduce. Another way to state this is that the genetic makeup of a population reflects the results of natural selection; natural selection is the name given to the observation that traits that increase the likelihood to survive and reproduce become more prevalent in that environment. If the environment is stable over time, then the genetic endowment evolves to fit that environment. Under dramatic changes in environment, a genetic endowment that is very dependent on those elements of the environment that have changed will be less likely to express that genetic trait in its fullest range.

The study of a broad field such as ecology also reveals that changes can occur within systems that do not affect other elements of that system. For example, a change in one or even several species might have no effect on other species and therefore no effect on the system as a whole. This restates one of the central ideas of the Aristotelian approach to causality: the scale at which an analysis takes place significantly influences the results of that analysis. Some causal mechanisms can only be discerned by observing the relationship between two elements. Others require examination of the interactions among many elements, and still others can be observed and studied only at the system level.

Thus, Levin's observation that "a full picture requires integrating and matching explanations across scales. . . . Patterns at certain scales are largely imposed, while those at other scales represent self-organization" is reminiscent of the Aristotelian approach to the programmatic level of causal analysis and is applicable to the study of systems in general.

However, are these true causal mechanisms? No. Rather, if corroborated and validated, they might be in the same category as Newton's laws or the laws of thermodynamics. They seek to describe a starting point, a set of principles from which causal actions can be derived. This is not to suggest, however, that the causal mechanisms underlying them cannot be explicated in the future. Newton's laws are now seen to be derivable from relativity theory, for example. If these or other generalizations about ecological systems do hold up to scrutiny and are widely accepted, they can then serve as a set of basic principles from which causal mechanisms can be derived. If they are not generalizable across all ecological systems, are wrong, or can be derived from more basic principles, then they cannot be a starting point from which causal predictions can be derived. Thus, it is too early to know whether top-down principles exist that can guide the identification of causal mechanisms operating at the level of ecological systems. The relatively young discipline of ecology is thus an example of a subject in which causal mechanisms at the programmatic level may be identifiable, but it is too early to know whether this is the case.

TIME AND BIOLOGY

Chapter 1 posited the existence of unidirectional time as a necessary precondition for causal reasoning, but chapter 6 noted that time would go in the opposite direction when moving at above the speed of light. Biological time has several features that meet the unidirectional, sequential requirement needed to establish causality. First, it is directional, because many biological interactions at the molecular and macromolecular level are irreversible and cannot be "undone" once that interaction has occurred. For example, once a nerve has fired, the chemical and electrical events that have taken place cannot be reversed, even though the chemicals that are released in the firing are recycled. This irreversibility is especially true of the many biological processes that are multistep sequences that cannot run in "reverse" order; even when individual steps are reversible, the chain of events contains many steps that are not. As a result, many biological processes ranging from molecular protein-protein interactions to the attainment of adulthood consist of a chain of events taking place in a specific order that can be known. Since causality requires that a cause occur before its effect, biology provides many examples in which this necessary element of determining causality is present.

A second element of biological time is the presence of innate "clocks" in many organisms. These are made up of sequences of molecular interactions that take place over a specific time period, that are repeated in a regular fashion, and that organize behavior in a predictable, regular fashion. Two examples familiar to humans are the circadian (Latin for "around the day") clock, which runs on a cycle of approximately twenty-four hours in length and expresses itself in the regular sleep-wake cycle that most people follow, and the menstrual cycle, which is approximately one month in duration and controls ovulation and the sloughing of the uterine wall (menstruation) in women of childbearing age. The study of these internal clocks and the regular patterns associated with them are the province of the field of chronobiology.

The primary director of the circadian rhythms in humans is a small group of cells in a brain structure called the suprachiasmatic nucleus (SCN). These cells function as a single clock that directs a number of different processes that are repeated in a predictable pattern. The period of this clock varies slightly from person to person (it averages 24.1 hours) and it is "reset" by the light-dark cycle (the environment). The clock can be forced to reset rapidly by traveling across more than one time zone; jet lag is the physiological consequences of suddenly disrupting the usual pattern of input provided by the regularity of the light-dark cycle.

The SCN regulates multiple processes besides the sleep-wake cycle, among them a predictable pattern of body temperature variation throughout the day, which is usually lowest several hours before awakening, and a patterned secretion of the hormone cortisol, which peaks daily on arising and is lowest around 10 p.m. These cycles result from the coordinated, sequential interaction of a specific series of proteins located in multiple areas of the brain and, when the function being regulated is located in another organ, a site remote from the brain. A second clock, located in an area of the brain called the hypothalamus, is entrained by eating and regulates other activities. Other clocks are located outside the brain and regulate still other activities.

A third aspect of biological time is the sequence of events that occur over the lifetime of a single organism at specific, predestined points in the life cycle of that individual organism. These are referred to as developmental processes and usually occur only once. An example is the onset of puberty in humans. The developmental sequence starts at conception and is directed by genetic programs that are themselves "built into" the genome of each species.

Nongenetic environmental events can have major effects on this sequence and have a meaningful impact on the ultimate makeup of that individual. Take the example of height. Individual humans achieve their final (adult) height over time but not at a constant rate. An insufficient exposure to a necessary element such as a specific nutritional substrate or sunlight needed to produce vitamin D or an exposure to environmental toxins that interfere with the innate developmental process at crucial times during fetal life, infancy, or adolescence can significantly lessen achieved height and thwart the (genetic) developmental program in an irreversible fashion. The result is a permanent deviation from its predetermined outcome. Normal brain development offers another example. Many cells in the brain form and then are "removed" or pruned in specific sequences as the development of the fetus and infant proceeds. Environmental events such as the taking of a single-dose medication that interferes with this process can result in abnormal development and lead to permanent dysfunction or death of the individual if the exposure occurs at a crucial time. Normal development thus results from both innate, predetermined sequences of events and from the interaction between the environment and these built-in tendencies. Since developmental events must occur in a specific sequence and are often irreversible once they occur, subsequent events are dependent on the initiation or completion of prior events.

These examples illustrate one other aspect of biological time: the heightened significance of specific time periods during the developmental process. These relatively brief but crucial time periods are reminiscent of the "nodal points" identified by systems theorists and discussed in chapter 5—structures, episodes, or incidents that are connected to many other events that, as a result of their high connectivity, play an essential or strategic role in causal sequencing. This heightened importance of specific events or time periods illustrates a mechanism by which nonlinear causality occurs in biology.

So far, this discussion of time in biological causality has emphasized innate processes, but the previous discussion of infectious disease epidemics demonstrates that unidirectional time also functions in more macroscopic environmental sequences. Many regularly recurring epidemics such as polio require the accumulation in the population of a large enough number of vulnerable individuals to allow the organism to spread readily both within population groups and between groups. Once a large enough number of individuals has been exposed and become immunized or died, the number of individuals vulnerable to the disease drops, and the epidemic resolves. The reaccumulation of enough vulnerable individuals to support once again rapid, widespread new infection requires time. Influenza epidemics are a variation on this theme, since they are caused by the emergence of new strains of the virus. The influenza virus is genetically predisposed to mutate (a predisposition that is a part of the causal chain), and it continually mutates in susceptible species of fowl, pigs, and humans. As a result, the likelihood that a mutation will occur that is both different enough from those previously in existence that prior immunity will not protect and sufficiently toxic to cause severe disease increases over time. Here, the passage of time increases the likelihood that a chance event, a mutation that causes disease in humans, will occur. Thus, while time plays a causal role in epidemics of influenza and polio, it does so in a passive, probabilistic fashion. This is an example of time as a necessary element of the programmatic causal sequence.

In summary, the ubiquity of unidirectional biological clocks and irreversible developmental sequences throughout nature in both plants and animals is an indication of the centrality of directional time to biological function. Perhaps it is one source of the human conviction of the existence of the construct of causality.

APPLYING THE THREE-FACET MODEL TO BIOLOGY

From a top-down view, causality in biological systems can be understood as acting at multiple levels simultaneously (facet 2), much in the way that Aristotle suggested. For example, some elements of the genetic program can be considered *predisposing* in any individual because they contain the directions for likely but not absolutely predetermined outcomes. This sequence can be interrupted by *precipitating* events such as an exposure to hormones in the water supply or prescribed medications or deprivation of needed nutrients at specific times that can have permanent effects on the expression of this aspect of development. Elements of the genetic program can also be precipitating. For example, all individuals who inherit one copy of the gene that causes Huntington disease will develop that illness if they live long enough. Puberty is an example of *programmatic* causality. It results from the occurrence of many different events and a multitude of processes that are orchestrated by multiple genetic programs and influenced by multiple environmental events. This process of orchestrating the many events that must occur in sequence can be conceptualized as multiple parallel and multiple sequential programs that initiate, maintain, and stop many prescribed, that is, programmed, events. The "programmer" and the orchestrator is the genetic code and in that sense represents programmatic causality, but the whole sequence can be conceptualized as more than the individual parts, an example of emergent causality.

From a bottom-up point of view, each element of the sequence of molecular events that underlies a particular biological process, for example, the production or inhibition of a specific protein product at a specific time, the opening up of a receptor that causes a change in the conformation (shape) of a protein, or the disabling or dismantling of a protein, brings about a specific molecular event that, if incorrectly carried out, can result in a *categorical* abnormality such as a birth defect or death. Many events, for example, final achieved height, depend on graded, *dimensional* differences in the amount of a substrate that is available or the amount of time over which it is active. These result in the bell-shaped distribution of universal characteristics and are best explained by probabilistic causal modeling. *Nonlinear* modeling helps understand the cause of interfering with developmental periods. For example, at the end of World War II there was a severe famine in the Netherlands. The deprivation of nutrition had effects on fetuses that they carried throughout life, including shorter achieved height as a group and an increased risk of schizophrenia.

Although these examples of biological time contradict relativity theory's prediction that time is reversible, they do not contradict the other central feature of relativity theory: the rate at which time proceeds depends on the relationship between the observer and the observed. Biological clocks can be "speeded up" or "slowed down" by changing the relationship of the organism to the driver of the clock. For example, altering the day/night cycle by flying across time zones results in the discomfort of jet lag. This resolves when the innate clock is reset by the dark/light cycle at the new location. The resetting occurs by "speeding up" or "slowing down" the innate clock so that it matches the external driver, the dark/light cycle at the new site. A trip of similar length in which an organism travels the same distance in a north-south direction but stays within the same time zone does not result in a changing of the length of the day because the light/dark cycle is not dramatically altered. Similarly, the preprogrammed events of life-course development can be caused to occur earlier or later by altering events or exposures that are necessary for that event to occur. An example of a positive adaptation to relative time is the use by some organisms of the unpredictability of a necessary event to regulate the timing of their life cycle. The Dauer (German for "permanent") larvae of some species require water to mature or procreate. Species that live in areas in which water availability is unpredictable and infrequent can go into a state in which their biological processes function very minimally for years. They appear to be in "suspended animation." When it eventually rains, the organisms quickly resume development and reproduction.

Might cloning provide a counterargument to the claim that biological time is directional? Thus far, the answer is no, because cloning—the removal of a stem cell from an adult organism, its placement in an enucleated egg, and the provision of the support it needs to develop into another adult—has not resulted in an *absolute* replication of the organism. Whether it is the operation of chance (discussed below in the next section), an inability to replicate exactly the environment in which it develops (perhaps a manifestation of the uncertainty principle), subtle alterations induced by the process of nucleus removal, postformation changes in the structure of the gene, or a combination of several or all of these, all cloned organisms to date have been subtly different from the organism in which the stem cell was obtained, just as "identical twins" can be recognized as distinct by those who know them. Whether this will remain true in the future is unclear, so it remains possible that cloning will be an example of a biological "resetting" of the directional time clock, but this is not the same as reversibility.

Finally, directionality in biological systems is an essential feature of the overriding theory of biology, Darwinian evolution. Evolution occurs over time in a specific sequence that is neither predetermined nor repeatable. Similar biological features have emerged several times over the course of the Earth's aging, supporting the notion that something, presumably the environment, is "choosing" or influencing this emergence, but there is no evidence this author is aware of that an exact replica has evolved more than once. A discussion of evolution appears in chapter 11.

This discussion of biological time also illustrates the application of facet 3's *empirical* logic. Experimental evidence from an extraordinary range of studies supports the ubiquity of biological clocks, their directional and irreversible nature, their cyclicity, and their molecular and macromolecular mechanisms. The claim that any process so ubiquitous must have some causal explanation would not meet Popper's strict criterion of falsifiability, but their ubiquity across the plant and animal kingdoms and the multiplicity of mechanisms that exist to entrain daily, seasonal, annual, and even longer periods of cyclical activity suggest something more. This is the appeal of facet 3's *narrative* logic. One might decide not to speculate on why biological clocks are ubiquitous, but avoiding the question or claiming that its inaccessibility to hypothesis-driven experiment makes it "unscientific" is choosing to ignore or overlook a fascinating and obvious question. Clocks allow the organism to function at a higher degree of readiness at specific times and low or no degree of readiness at others. Animals and plants most active in the daytime function best in the sunlight and have physiological functions such as higher cortisol levels and maximal degrees of alertness at that time of the day. For nocturnal organisms, the opposite is true. Invoking a Darwinian explanation—that these functions coexist because they increase the likelihood that the organism will survive to reproduce—offers a quite powerful, albeit untestable, explanation. I will argue in chapter 11 that the concept of evolution is so powerful *because* it provides a narrative that can be applied in many such circumstances and because it is impossible to refute. Undoubtedly, some will disagree and conceptualize the existence of natural selection as empirical, but the existence of biological time begs an explanation. The creation story in Genesis, in which God established the existence of day and night as the first step in the creation of the universe, speaks both to the centrality of this mystery and to the use of *ecclesiastic* logic as another causal approach.

Does the seeming contradiction between modern physics' conception of time as nondirectional and the claim made here that biological time is unidirectional mean that one must be wrong? I suggest that these two quite distinct models of time demonstrate how a construct can have different, even contradictory, implications or interpretations when viewed from different levels of analysis. In modern physics, the subatomic view of time predominates, even though cosmologists are interested in the evolution of the universe over immensely large spans of time, and though events at or even faster than the speed of light are plausible. In biology, the study of life on planet Earth, the organismic view predominates, even though the study of molecular-level phenomena and even their subatomic bases are very active areas of inquiry, and though the ubiquity of clocks and the irreversibility of many biochemical pathways and events make unidirectional time determinative. This acceptance of seeming contradictory views is reminiscent of the wave/particle duality—both its claim that matter can exist simultaneously in two ostensibly different forms and its claim that it is the observer who determines which property is emphasized.

A similar proposal underlies the approach to causality proposed in this book. If one posits the existence of causality and justifies this assumption by the benefits that accrue from its application, that is, its utility, then the directional model that underlies much of this book is justified. Unidirectional time best explains and is most compatible with the world as it is known to biologists: events have causes that can be discerned, and methods can be used to support independently the cause/effect relationship. On the other hand, beginning with the premise that the speed of light can be exceeded and that time is therefore nondirectional also leads to many accurate predictions and thus has both utility and beauty. Requiring that one view be more powerful or "truer" than the other would lead to a significant loss of explanatory power. That is, the benefit of accepting these seemingly contradictory ideas is that the power of explanation is maximized.

CHANCE IN BIOLOGICAL CAUSALITY

The role that chance plays in understanding causality is an issue touched upon elsewhere in this book but not discussed explicitly. In a fascinating, well-written book, *Chance, Development, and Aging*, Caleb Finch and Tom Kirkwood suggest that chance is one of three elements, in addition to genetics and environment, that play meaningful roles in the development of organisms. In their view, the word "chance" has two meanings in biological systems. "Random chance" refers to outcomes that are totally unpredictable, while "stochastic chance" refers to situations in which variation in outcomes is predictable. Two examples of stochastic variation in gene expression are the exact time it takes for the protein produced by a gene to interact with its specific receptor and the variability in the distance that a substance must travel (diffuse) before reaching its target. The variation in each of these is constrained within limits and can be mathematically described. Furthermore, this variation can result in different outcomes. Because stochastic chance can be quantified in this way, it can be incorporated into causal models.

While the likelihood of random chance can also be mathematically described, it is unpredictable and therefore different causally. That is, stochastic chance is a form of probabilistic or dimensional causality and is more likely to result in minor variations than distinct dichotomous outcomes. As such, it is more akin to *predisposing* causality. Conversely, random chance is usually unpredictable and more likely *categorical* and *precipitating*. However, there are exceptions to these generalizations, for example, when the timing of an event or the amount of a substrate is so tightly linked to subsequent events that a minor variation leads to a major developmental abnormality. This would be an example of *programmatic* causality if it resulted from an effect on a highly connected gene or protein.

Three examples of chance that follow the stochastic model are the selection of which of a group of identical cells migrate and which are pruned, which half of an asymmetric cell division develops, and

the Brownian motion of molecules. Each can result in different outcomes and thus be causal, but the occurrence of each is predictable
in a given system when enough observations are made. Examples of
random chance that are causal are exposure to a toxin at a specific
point in development and mutations induced by cosmic rays from
space. Stochastic chance is an essential aspect of the process; random
chance is externally imposed on the process.

Thus, stochastic chance is somewhat akin to the uncertainty principle; it places a limit on the absolute accuracy of measurement and
prediction and, thus, causal prediction, but it in no way eliminates
the ability to make extraordinarily accurate predictions. In Finch and
Kirkwood's view, study and experiment can illuminate and identify
the mechanisms underlying chance variation in biological systems
and, by describing and constraining the limits of variability, improve
the accuracy of predictions. For example, in a discussion reminiscent of the centrality of initial circumstances in chaos theory, they
note that the requirement in many biological processes of a given
amount of a substrate at a specific time can lead to a dramatically different later outcome when small differences in amount are present
at the beginning of the process. By examining many examples of this
event, one could predict the likelihood of specific outcomes, and, by
identifying factors that influence the availability of that substrate at
that specific time, understand the stochastic differences in outcome.
However, like the uncertainty principle, stochastic chance cannot
predict the outcome of an individual instance, because that can only
be known after the event has occurred. Stochastic chance indicates
that variation is inherent and unavoidable.

By offering an approach to studying and codifying the mechanisms of chance, Finch and Kirkwood offer an antidote to the generally held idea that all chance is synonymous with total unpredictability. For example, the recognition that two specific enzymes in
the influenza virus have the ability to mutate frequently and that
this leads to the constant emergence of new strains of the virus has
led to surveillance programs that monitor the virus worldwide and
identify which new strains are particularly virulent. This knowledge spurred the development of procedures by which vaccines that
target new variants can be mass produced and distributed within

months, thereby greatly diminishing the likelihood of an epidemic. Thus, the recognition of the mechanisms by which stochastic chance operates led to a set of interventions that have minimized an undesirable outcome.

It is interesting to note that Darwin speculated that chance variation is a means by which natural selection operates. Prediction remains possible, but not at the level of an individual organism. That is, chance does not eliminate prediction but constrains it.

Finally, Finch and Kirkwood remind us that the language used to describe such processes can influence how they are interpreted. Emphasizing the randomness of chance might imply to some that it is not amenable to study or measurement, that the events are unimportant, or that they are beyond the study of science. None of these is true. Furthermore, attributing events to chance can influence the value ascribed to them. For example, environmental events occurring at crucial times during development can be described as chance events or as environmental influences. The word "chance" implies that they cannot be avoided or that causal attribution cannot be made, while the word "influences" implies that they are amenable to influence and understood as causal in nature. This example of how the meaning of words can carry very different causal implications even when describing the same circumstance is an illustration of the importance of narrative, a topic to be discussed in chapter 9.

GENERALIZATIONS ABOUT CAUSALITY IN THE BIOLOGICAL REALM

1. Causality in biological systems is often complex, but not indeterminate. Many of the examples discussed in this chapter demonstrate the complexity of causal relationships in the biological realm. This is because many events of interest (adult height, sinusitis, a heart rhythm disturbance causing death) result from a set of interactions among multiple factors. This should not be surprising given the complexity of many normal physiologic events such as the immune response to foreign invaders, electrical conduction in the heart, and the achievement of height in an individual, but this knowledge should caution us to be careful when considering phrases with causal

implications such as "nature versus nurture" or "genetic versus environmental." Simplicity may be beauty in the eyes of some (à la Occam's Razor), but what one calls simple is a judgment by which nature does not always abide. Nonetheless, scientific study over the past several hundred years clearly demonstrates that causal explanations can be identified.

2. Analysis at multiple levels maximizes explanatory power. While Galileo was right to observe that the complexity and difficulty in operationalizing the Aristotelian model limited the ability to study causality in nature, many of the tools and methods developed in the centuries since Galileo have enabled scientists to identify and characterize multiple causal factors. As several of the examples in this chapter illustrate, a detailed examination of causal relationships at multiple levels is now both possible and fruitful. For example, the tools of the molecular biologist, the population geneticist, the infectious disease specialist, the epidemiologist, and the health planner were necessary to eliminate smallpox from the earth because they were able to identify the causal steps at which intervention needed to be targeted. Galileo was not wrong. Rather, this is the first time in human history that the methodological and conceptual tools needed to understand multiple agents operating at multiple levels are available.

3. Biological time has inherent, irreversible directionality.

4. There will always be limitations in the ability to predict outcomes in biological systems, but accuracy in prediction can increase with further study. The relative uniqueness of individual events, whether it is at the level of the molecule or the environment, combined with the complexity of many causal relationships limit the ability to make accurate predictions in any single instance. The challenge in many biological examples is not to identify a single causative agent but rather to determine how much causal influence should be attributed to each of the interacting elements. This is another way of noting the probabilistic or dimensional nature of many causal relationships. This biological uncertainty principle limits absolute certainty because all parameters cannot be specified with certainty, but it is clear that broad generalizations about causal mechanisms can be discovered.

Limits on the ability to describe causal relationships in biological systems result from the large number of elements in many systems, the interaction of many different systems at different levels, the role played by chance events, and a disproportionate importance of initial (early) conditions or events.

The use of multiple methods and multiple viewpoints increases the ability to identify and understand specific causal elements, and continued improvement in methods and measurement are likely to improve further the ability to identify causal mechanisms and outcomes. Thus, even though the "biological uncertainty principle" means that accuracy can never reach 100 percent, further study is likely to increase both knowledge and improve the ability to make predictions.

5. Replication, falsifiability, and hypothesis testing are important tools for demonstrating accuracy and causality in biology, but there are inherent limits to their application. The ability to repeat and replicate what has happened in nature is limited. This places constraints on the use of the tools of replication, falsifiability, and hypothesis testing to study causal mechanisms. Every species, every environment, every system, and every universe has elements unique to it. Much of the gain in scientific knowledge in the past half millennium has been accomplished by the approach Galileo recommended— isolating specific elements for study. A major challenge for the future is to develop methods that can better test proposed causal mechanisms at the systems level.

GENERALIZATIONS IN THE SEARCH FOR CAUSAL MECHANISMS IN LARGE SYSTEMS

This chapter also suggests broad generalizations about causality at the programmatic level:

1. The definitive identification of all causal influences is not possible in large, complex systems. Because large systems are made up of many elements, it is not possible to describe *simultaneously* all the relationships among the constituent elements with completeness

and mathematical precision. Furthermore, systems are not static but rather change over time. A change in the relationship between just two elements of a system is likely to result in changes to other aspects of the system with which they interact. Said another way, the relationships among the individual elements of the system are themselves complex. As a result, it is not possible to specify with exactitude all the causal contributors and the degrees of their contribution to causality in complex systems.

These two issues, the limited ability to identify all interactions and the impossibility of quantifying them simultaneously, also limit the ability to predict the outcome of a change in one part of the system. This is especially true over the long run, since chaos theory predicts that small initial differences can have major effects on long-term outcome.

Thus, many systems cannot be completely described, in part because the notion and boundaries of a system are imposed by humans (point 1 in Levin's list) and in part because nature is organized such that the number of interactions increases exponentially as the number of elements increases, so that it is impossible to describe every element of that system. These limitations are reminiscent of both Gödel's incompleteness theorem and Heisenberg's uncertainty principle.

2. Complex systems develop or incorporate mechanisms that limit or encourage change. The structure of the water molecule, for example, influences the effect of temperature change in many abiotic and biotic systems. In the biotic world, a number of genetic mechanisms have developed that eliminate "unexpected" DNA sequences or their encoded proteins and so lessen an individual organism's likelihood of survival when such a change occurs. These mechanisms can be conceptualized as having the causal purpose of constraining or encouraging certain causal outcomes. Thus, as Levin and Aristotle noted, systems develop mechanisms that appear to have causal attributes at the system level but that emerge from unique elements of individual or relatively constrained elements of the system.

3. The establishment of the boundaries of complex systems is arbitrary. For example, if one is interested in explaining the changes

in the Earth's atmosphere that have occurred over time, does one include the Earth's core and the sun in the system, since both are important sources of energy? And how about the space between the sun and the Earth, since it too has an influence on the constitution of the atmosphere? Thus, boundaries are often arbitrarily established to allow study and discussion to proceed. On the other hand, there are discontinuities in nature that isolate separate systems, Earth from the rest of the cosmos, for example, and provide convenient limits within which the search for causal mechanisms can be constrained. In addition, the definition of a system might evolve as relationships among elements are discovered or be influenced by the specific topic that is being studied.

DIFFERENCES BETWEEN THE BIOLOGICAL AND PHYSICAL SCIENCES

The physical sciences and biological sciences face many of the same challenges in the search for causality. These include the need to explain phenomena at a range of levels from the very small (for example, how a change in electrical charge can open a channel that allows only a specific ion to enter a cell) to the very large (for example, why a single species of tree predominates in a geographical area of several thousand square miles) and a desire to extrapolate from currently observable phenomena to events that occurred long ago. This is not surprising, since the building blocks upon which the biological world is based are the same as those studied in the physical sciences.

Another similarity between these two broad disciplines is that both organize the material they study into multitiered, interacting systems —biology builds disciplines around such constructs as molecules, cells, organs, functional systems, organisms, and species, and physics examines matter at such levels as paired subatomic particles, atomic nuclei, elements, rocks, tectonic plates, planets, solar systems, etc.

However, there is a major difference between biology and the physical sciences of chemistry, physics, and geology, and this difference has a significant impact on the topic of this book: we are more likely to assign teleological or causal purposes to these different levels in biology than in physics. For example, biologists who study ion channels in nerve cells recognize that these cells are part of a system that propagates electrical messages along groups of cells and that each organized pathway ends in one or more specific collections of central nervous system cells that have identifiable, often unique, functions, such as sensation. Biologists, however, go on to attribute *purpose* to these sensations, among them the avoidance of danger (pain pathways, for example) and the identification of something to eat (smell and vision pathways). Biologists (and people in general) also assign purposes to groups of organisms such as colonies, families, herds, tribes, cities, states, and nations. For example, certain colonies of bacteria, when acting together, secrete organic materials that help them reproduce or capture nutrients; the amount produced by any one organism could not support its survival on its own, but when many organisms produce the material simultaneously, the group and thereby the individual organism benefits. Similarly, human families provide physical and emotional support that maximizes the development, survival, and perpetuation of subsequent generations; groups of families working together allow for protection from predators and the distribution of work; and larger groups yet—communities, states, nations, and groups of nations—further increase the likelihood that their members will survive and thus propagate the genes and ideas of both the individuals and the group. Thus, we attribute purpose to each level of interaction, and purpose sometimes implies causation. This does not occur in the physical sciences. Sometimes these causal attributions are also applied with little appreciation that they are being given such a meaning. For example, species are described as adapting to new niches, extending their range, or adapting to dramatic changes in predators, climate, or geography; each of these implies a purpose and thus a cause for an observed activity.

This is not to deny that there is a human-imposed aspect to the description of different levels of analysis in both biology and physics or to claim that the distinctions made about different forms of matter by physical scientists are invalid or unimportant. Rather, this is a claim that the parsing of biological functions into different levels of analysis can be influenced by observations about causal actions occurring at these different levels of biological organization, whereas in the physical sciences, the analysis of phenomena at different levels reflects either the range of maximal function of an instrument or an inherent characteristic of nature. The search by many physicists for a "unified field theory" or "theory of everything" that will explain how matter functions at all levels of analysis reflects their belief that a single coherent explanation of matter at all levels of analysis is possible, something that is not true of the biological sciences.

Should this tendency to attribute purpose to causal reasoning in the biological sciences be criticized and avoided? Galileo suggested that this fourth level of Aristotelian cause be put aside. He was focused on the physical sciences, and, still today, purposive explanation seems to have little if any role in causal explanation. However, purposive explanation continues to be widely applied in the biological sciences. I conclude that this demonstrates the power of causal purpose to capture the human imagination and that it would be foolish to avoid or abandon it completely. Rather, using the Aristotelian model to separate it from other aspects of causal reasoning seems most likely to minimize the risks of applying it inappropriately while allowing continued study of many of the "big" questions of the "why" of things.

8

EMPIRICAL: EPIDEMIOLOGY

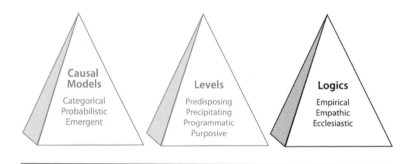

Epidemiology is the science that uses the study of populations and groups of individuals to identify correlates and causes of ill health. Its goal is the identification of strategies to prevent and treat disease. The discipline developed in the mid-nineteenth and early twentieth centuries out of a desire to analyze accurately vital statistics such as birth and death records being collected by governments and from the recognition that some diseases could be prevented by targeting groups of people rather than individuals.

The word "epidemiology" shares a common root with the word "epidemic," and Hippocrates is credited with distinguishing between the episodic nature of epidemic disease and the persistent nature of endemic disease. In the fifteenth and sixteenth centuries, epidemics of the Black Plague killed widely in Europe and were greatly feared, and today epidemics of influenza, gastrointestinal illness such as norovirus, SARS, and HIV/AIDS are frequent newsmakers.

Around 1600, the Borough of London began collecting weekly counts of christenings and burials in order to assess the status of the plague. The data were little used, however, until 1854, when John Snow, who was working for the borough, began mapping where people lived who had developed a sometimes fatal diarrheal illness that we now know was cholera. Snow noted that cases of this illness clustered in certain neighborhoods and that these clusters centered on the water pumps from which those communities drew their water. He concluded that the diarrhea was propagated by fouled water and hypothesized that the epidemic could be halted if neighborhood residents were barred from using the source of the water. Famously, Snow disabled the handle of the Broad Street water pump with a chain—and the diarrhea outbreak resolved.

What Snow did was to use the careful collections of statistics to identify an association between a specific illness in a particular group of people and a shared characteristic of that group (geography in this example), identify what these people had in common (exposure to the water pump at the center of the geographic area), hypothesize a link between that exposure and the spread of the disease, and then intervene to prevent further cases of the epidemic by interrupting this hypothesized mode of transmission. This multistep reasoning process is a nice example of applied causal logic, but, even today, some scholars argue that the outbreak was already abating by the time Snow disabled the pump handle and that his action was not the actual cause of the epidemic's resolution. Nonetheless, no one doubts the causal link he identified between the water source and the epidemic, since cholera is spread by the ingestion of water that contains the cholera bacillus.

Today, epidemiological studies examine risk factors such as diet, behavior, genetics, workplace exposure, geography, and food preparation practice and seek to identify associations between these risk factors and specific endemic and epidemic health concerns. But such studies face the same challenges that scholars raise about Snow's finding of an association. How can one be sure that an association between two events is a causal one?

$$\Delta \quad \Delta \quad \Delta$$

Interestingly, Snow carried out his experiment before the discovery of the germ theory of disease. Like Jenner, who a century before had discovered that inoculation with material from cowpox pustules could prevent smallpox, Snow was credited with averting the development of future cases by preventing transmission without knowing the identity of the causative agent.

Through the application of complex causal logic, Jenner's discovery of vaccination led to one of the great triumphs of modern medicine, the total eradication of the disease smallpox by 1980. This effort was carried out by the World Health Organization and depended on the hypothesis that the transmission or spread of an infectious disease could be halted if most people in the population are immune to the causative agent. This hypothesis is based on the observation that once people are infected with a specific agent, a virus in the case of smallpox, they cannot be reinfected. Thus, a disease caused by an infectious agent can continue to afflict humans by only four mechanisms: continually "finding" new, never infected individuals; infecting and surviving in other species until never-infected humans "become available"; surviving in a rare human who does not develop immunity; or surviving on inanimate objects until never-infected humans "become available" for infection.

Since the smallpox virus survives only in humans, it must maintain viability by infecting vulnerable individuals. It is a very contagious virus and spreads quickly among those who are not immune. In the past, the virus would sweep through a population and infect almost every never-exposed individual. Thereafter, it would be maintained in "reservoirs" of small isolated populations, usually in remote parts of the world. After several years, a large number of never-infected children would be present in more populated areas, and when the virus was reintroduced into these populations it would spread quickly among these children and rare uninfected adults in the pattern recognized as an epidemic.

Using straightforward causal logic, public health scientists reasoned or hypothesized that smallpox could be completely eliminated as a disease if most of the individuals residing in those small pockets of the world in which the virus was being maintained between epidemics could be vaccinated.

An aggressive campaign was launched that combined vaccinating children and never-infected adults in populous regions and identifying those remote reservoirs in which the virus was being maintained between epidemics. This required the identification of these remote reservoirs and the establishment of surveillance teams that could quickly detect new infections. When a new outbreak or even a single individual infection was identified, a team would be dispatched to vaccinate those in that area who were vulnerable. Successively fewer and fewer pockets of vulnerable individuals remained, until the smallpox virus was eliminated from the last human reservoir and thus the entire human population. Today, smallpox is no longer a scourge of humankind, and vaccination against smallpox is no longer necessary because the virus does not exist in any human. This is an extraordinary achievement that resulted from an understanding of the specifics of the causal web of that disease. While an ever increasing number of never-vaccinated individuals is susceptible to smallpox, the disease will recur as a human scourge only if it is purposefully reintroduced by humans with diabolical intent.

Cholera, in contrast, remains a devastating infectious disease that still causes epidemics of illness 150 years after Snow's intervention demonstrated that its spread can be interrupted if infected water supplies are eliminated. The cholera bacterium is known to be spread by the fecal-oral route, and sewage treatment and drinking-water chlorination have limited or eliminated it from many developed countries. However, water and sewage treatment plants are expensive and require long-term investment, and even today they are not available in many parts of the world. Furthermore, the cholera bacillus is maintained in nonhuman reservoirs, and its biology has so far thwarted development of an effective vaccine. Thus, even when knowledge of the biology and epidemiology of different agents is comparable, the specifics of the causative agent have a major influence on what can be done to break the causal chain.

The Broad Street pump incident, the elimination of smallpox, and the failure to control cholera in parts of the world have several important lessons to teach. First, even though knowledge of the biology of the smallpox virus and the cholera bacillus has contributed to the elimination of smallpox and the control of cholera in some places, the smallpox vaccine was developed and the pump handle chained long before the causal agents were identified—even before the concepts of virus and bacteria were developed. Thus, outcomes can be changed even when knowledge of their cause or causes is limited or partial. Second, the elimination of smallpox and the local prevention of cholera required not only the identification of the causative organism but also an understanding of the dynamics underlying the spread of each specific infectious agent. These mechanisms of spread depend ultimately on the organisms' biology, of course, but they also depend on the environment external to the individual organism.

Third, the spread of these organisms and the epidemic nature of the illnesses they cause result in part from how these agents act in *groups* of individuals and how humans organize themselves independent of the organism. That is, human behaviors such as living in crowded settings, going to school, touching each other, and traveling from remote parts of the world to more populous areas are part of the causal web. Understanding them was and is crucial for designing and carrying out control and elimination strategies.

For example, smallpox elimination required not only methods for identifying those places in the world where cases were still occurring but also the cooperation of many governments and nongovernmental agencies, quick access by specially trained health care teams to isolated areas in several parts of the world, and the financial means to carry out a coordinated, multinational effort. Suspicious governments had to be convinced that the goal was worthy, and isolated peoples who had had little contact with the outside world needed to be courted and convinced that they and their children could benefit in order for the program to succeed. To do so, the program organizers had to learn about the values, social structure, and governance of multiple groups of often remote individuals, develop methods for introducing models of disease and disease prevention to people whose experience and worldview did not include modern ideas about these concepts, and learn how to train and use local individuals. Thus, the causal web of smallpox included much more than a causative organism with a specific method of spread. It included economic and political factors, patterns of residence and travel, values and social organization, and the very concept of prevention.

The causes of smallpox and cholera, then, are multiple and complex. They include a *provoking* cause, a specific infectious agent; a set of disparate *predisposing* causes that include characteristics of the human and nonhuman animal immune systems, patterns of contact between humans, patterns of contact between humans and animals, and hand-washing practices; and a set of *programmatic* causes that include the means by which water is obtained by a population, the economic capacity of states and regions, the availability of vaccination programs, the health beliefs of a populace, and whether those who control the political infrastructure allow access to groups of vulnerable individuals. Such information can guide successful programs such as the one that eliminated smallpox and can help pinpoint why programs to eliminate polio and cholera completely have thus far failed.

Another lesson that can be gleaned from the success of the smallpox elimination program is an important one: very complex causal patterns *can* be dissected, their constituent parts identified and understood, and the knowledge gained used to alter or eliminate events that are undesirable. While it took many decades to unravel the biology of the smallpox virus and several centuries to develop and implement a strategy that could address the biological, environmental, and governmental actions needed to unravel the causal web, the elimination of smallpox was accomplished relatively quickly after the strategy was proposed and the effort to carry it out implemented. This example offers a counterpoint to the concern that the complex nature of many causal webs makes the application of causal reasoning a futile exercise. Problems can be dissected and their complexity understood, and effective actions based upon these understandings can benefit humankind.

Several other successful public health interventions also focus on groups and populations rather than individuals and illuminate another point about programmatic causality. The fluoridation of water supplies and the dramatic drop in the rates of tooth caries and tooth loss, the supplementation of flour with vitamins that has helped eliminate certain vitamin deficiency diseases, and the addition of iodine to salt, which has greatly lessened the incidence of thyroid goiter, all depend upon "exposing" many individuals who would never have fallen ill to the prevention strategy. Like the soldier who does not fight but whose support function contributes to the war effort, these individuals are part of the causal web even though they might not individually benefit from the strategy that interrupts the disorder. That is, at the programmatic level of analysis they are elements of the system and thus parts of the causal web. Furthermore, the choice by a society to require that everyone be exposed to the prevention strategy rests on shared political and moral values such as the belief that a healthier population is one of the purposes of government. These shared values can be seen as an element of the programmatic level of analysis, as noted above. Chapter 10 will discuss the concept of cause in moral, spiritual, and religious contexts, referred to here as *purposive* cause, a fourth level of causal analysis. It is only noted in passing here that the goal of preventing suffering played a causal role in the choice of these prevention programs.

∆ ∆ ∆

While infectious diseases provide useful examples of the successful application of causal reasoning, other public health issues have proven to be even more complex and illustrate some of the challenges of causal reasoning. In the 1930s and 1940s, claims were made that cigarette smoking caused lung cancer. Like smallpox vaccination and the chaining of the Broad Street pump handle, the claim was made without knowledge of a plausible biological mechanism, and this, along with many other factors, fueled skepticism about a causal link. How, then, could it determined whether cigarette smoking is or is not a cause of cancer?

Koch's postulates, a categorical model of cause, was the operative model of cause when an association between cigarette smoking and cancer was first proposed. Could its three criteria—the identification of an agent that is always associated with the disease, a demonstration that the agent can induce the disease, and the elimination or diminution of the prevalence of the disease by elimination or control the causative agent—provide evidence to prove causality? In 1950, the British epidemiologists Richard Doll and Austin Bradford Hill demonstrated that the risk of developing lung cancer was significantly higher in those who smoked cigarettes than in those who did not. However, lung cancer occurred in people who did not smoke, and many individuals who did smoke had not developed lung cancer. Thus, Koch's first two postulates could not be upheld. And of course, Koch's third postulate could not be tested because it would be both unethical and impractical to expose some individuals to tobacco smoke and others to a placebo to determine if smokers were more likely to develop cancer. These challenges, as well as questions such as those raised about the effectiveness of Snow's chaining of the pump, led several prominent epidemiologists in the 1950s and 1960s to consider what kinds of evidence would be needed to establish causality.

CRITERIA OF BRADFORD HILL

- Consistency
- Strength of association
- Temporal sequence
- Dose response or gradient relationship
- Specificity
- Coherence
- Biological plausibility

The best-known and most widely cited criteria were developed by Bradford Hill (1897–1991) in 1954 and are listed here. The length of the list alone illustrates the challenge of convincingly demonstrating causality. It suggests that no single criterion or even several criteria (a term he never used) will be useful in all situations. Rather, as both Bacon and Hume had suggested centuries earlier, this approach suggests that many causal hypotheses are made plausible only by the development of multiple lines of evidence and, furthermore, that the greater the number of disparate lines of evidence, the stronger the case of a causal relationship and the lower the probability that an identified association has occurred by chance. Epidemiologists use the phrase *convergent validity* to describe this use of multiple lines of evidence supporting a single conclusion and the term *reliability* to describe multiple replications of the same experiment or result. Both approaches increase the likelihood that chance is not the explanation for the association found by a single study, but, again, convergent validity cannot prove with absolute certainty that an association is causal for several reasons: some other factor could still be the cause of the hypothesized antecedent, study and measurement themselves affect ultimate outcome, and because, as Hume noted more than two hundred years ago, inference is still required to make the final connection.

Another criterion that can support but not prove causality is reflected in the last of Koch's postulates: if interrupting the relationship between the alleged cause and the disorder prevents the outcome from occurring, a causal relationship is supported. As we have seen, though, the resolution of the cholera epidemic following Snow's chaining of the pump handle convinced many that his causal linkage of the water supply and the diarrhea was correct, but some modern-day historians believe that the epidemic was already on the wane—that is, the chaining of the handle had little or nothing to do with the subsequent resolution. The fact that lung cancer rates have fallen in the United States among men in parallel with declining rates of smoking in men strengthens support for the causal linkage between the two, but it remains possible that some other environmental change, for example, declining air pollution thanks to antipollution laws, explains some or all of the decline. This particular alternative explanation seems unlikely, though, since lung cancer rates in U.S. women have risen following an increase in the rates of smoking among women in the past fifty years. While some other explanation for this *dissociation* between lung cancer rates among men and women is still a possibility, this finding further significantly strengthens the causal claim.

Although the linkage between exposure to cigarette smoke and lung cancer is now widely held to be true, the biological mechanism remains to be discovered. The wide acceptance of the causal link between tobacco use and cancer likely reflects the strength of the convergent evidence (data are available that meet most of the types of evidence listed by Bradford Hill, the conviction among experts and medical practitioners that such a causal link exists, and finally the conviction among much of the public that smoking causes lung cancer). However, bringing about (causing) lower rates of tobacco use at the population level required more than convincing the public of the causal link between cancer and cigarette smoke exposure. It also required changing behavior by restricting access to and discouraging the use of tobacco products through legislation that bans sales to young people and significantly limits smoking in public places. The successful lowering of rates of tobacco use is a good example of programmatic-level causal reasoning. I will argue in chapter 9 that rhetorical methods play a central role in convincing people of the truth of causal linkages in both science and narrative approaches such as history. It is important to mention here, though, that it took the employment of several rhetorical devices to convince scientists, legislators, public health officials, and the public at large of the truth of the causal linkage. Nonetheless, some individuals, especially young people, are either unconvinced that there is a link, unconcerned even if there is a link, driven by other reasons (a discussion of which is beyond the scope of this chapter and book), or unaware of the link and begin or continue to smoke. This suggests that the line between science and narrative is not always sharp, even if one agrees that such a distinction exists.

HOW CAN MEDICAL TREATMENTS BE SHOWN TO CAUSE BENEFIT OR HARM?

Medical ills have been treated with physical, medicinal, and psychological treatments for thousands of years. Many of these therapies address presumed causes of sickness and claim, as did Snow, that a recommended treatment caused an improvement. While skepticism about the efficacy of specific treatments has probably existed as long as treatment has been tried, it is only in the past sixty years that a scientifically based method for studying the efficacy of these interventions has been widely accepted. This is relevant to the general issue of causality because interventions are used as one test of possible causality in many fields.

The current standard for demonstrating that a proposed treatment "works" is the randomized clinical (or controlled) trial, universally referred to as an RCT. A discussion of how the RCT became the standard by which evidence for treatment efficacy is evaluated is worthy of discussion because it demonstrates that improvements in causal reasoning can occur and illustrates the point that even the methods for determining whether a causal relationship exists can develop in a haphazard and serendipitous manner.

In 1753, James Lind wanted to test the hypothesis that scurvy, which occurred commonly among British sailors, could be prevented by eating citrus fruit. Lind performed a study in which sailors on one ship were provided with citrus while those on another were given usual provisions. No sailors on the ship that received the citrus developed scurvy; some on the ship stocked with the usual provisions did. Having this nontreated comparison or *control group* allowed Lind to determine that the change (no scurvy) seen in the group exposed to the treatment (the citrus) could be ascribed to that treatment rather than to chance or some other unidentified element in the environment. That is, the citrus led to (caused) the prevention of scurvy and, by inference, provided some antidote to the development of the disease. The eventual provision of citrus fruit to British sailors led to their appellation as "Limeys," but little attention was paid to the general principle behind the method Lind used for almost two hundred years.

One of the challenges of proving the third step in Koch's causal criteria—that the introduction of the presumed causal organism led to the development of the disease—was identified by engineers studying factors that influenced the quality of light bulb manufacture at a General Electric plant in Hawthorne, New Jersey, in the 1920s. To their surprise, just observing the workers led to a change in the quality of light bulb production, that is, changed their behavior. While this seminal observation, which came to be known as the Hawthorne effect, was a nuisance factor for the engineers trying to identify procedures for improving quality, it also raised an important question about the determination of cause and effect: how can the introduction of a change be used to support the claim that it caused an effect when merely observing participants can change what occurs?

The answer was developed around the same time by the statistician Ronald Fisher, even though it would not be applied to clinical trials for another thirty years. Fisher, who developed many of the statistical tests commonly used today, had been hired to help the British agricultural service determine how to improve crop yields. For decades the service had been carrying out experiments in which the impact of different aspects of farming, such as watering, seed type, and soil types, were being compared, but it had run up against a problem of interpretation—since no two fields were exactly alike, how could it confidently attribute a difference in outcome between two different fields that were being "treated" differently to the farming practices that were being compared, when the fields might also differ in amount of rain or sun, soil type, or prior crop use?

Fisher's solution was twofold: first, compare many fields, and second, use random assignment to determine which fields underwent the "intervention" farming practice that was being studied and which underwent the usual farming practice. This approach accomplished two crucial goals. First, the comparison of a single intervention carried out in many fields to usual farming in many other fields lessened the likelihood that chance would account for any observed difference, since the larger the number of plots compared, the more likely it is that any real difference would be observed. Chance alone might account for a difference if a small number of fields were being compared, but would be unlikely to explain observed differences when many fields with one intervention were being compared to many fields in which the comparison approach was being carried out. Fisher suggested that a threshold be set as the standard for excluding chance as the explanation for an observed difference in outcome. While he never suggested a specific number, this is the origin of the now universal standard that a probability of less than one in twenty ($p < 05$) be used as the threshold below which chance is *un*likely to explain or be the cause of an observed difference. Comparing two approaches was not a new idea, as even Lind had studied several different potential treatments for scurvy, but establishing a required difference between the comparisons was revolutionary.

Fisher's second innovation was even more revolutionary but not fully appreciated for several decades. Scattering or randomly assigning the intervention among many fields had the effect of minimizing whatever differences naturally existed among the fields, since chance would make it likely that the differences that were not of interest would randomly and thereby equally exist in the two groups being compared. This required that systematic bias be avoided in choosing which sites would receive the intervention and which would not. Random assignment made it likely that variables *not* of interest—for example soil composition, weather, and watering techniques—would be "controlled for" or eliminated as sources of any difference between the plots receiving the usual practice and those receiving the new practice.

It was not until the late 1940s, when the drug streptomycin, an extract of a soil fungus, was developed as a potential treatment for tuberculosis, that Fisher's insights were applied to disease therapy. Tuberculosis, also called the White Plague, was known to cause the deaths of millions of people annually. Robert Koch had identified the causative agent in 1882 using the first two of his principles, but no treatment had emerged from this great triumph of the germ theory. The need to determine whether streptomycin could treat tuberculosis was clearly an important question, but streptomycin was difficult to manufacture and therefore in short supply. Hence, the team designing the trial, which included Austin Bradford Hill, wanted to study the smallest number of participants possible.

The researchers also reasoned, like Lind, that they needed to have a comparison group of untreated individuals, since some people with tuberculosis improved slowly on their own. At the same time, they wanted to avoid favoritism in assigning subjects to the streptomycin or untreated group, since tuberculosis was a fatal illness and because the treatment would take many months to complete. The answer they settled on was setting up a lottery to determine who would receive the active drug and who would not. This was a direct application of Fisher's concept of randomization to a treatment or control field, but it was fairness rather than control of chance differences that motivated this design element.

These innovations did not address the Hawthorne effect, however, so later studies added a third design innovation: exposing all individuals participating in the trial to exactly the same circumstances, thereby keeping them unaware of who is receiving the active treatment and who is receiving the comparison treatment. This masking or "blinding" assures that all individuals in a study have the same expectation that they might benefit. Blinding the researchers prevents them from unknowingly or knowingly treating those receiving the test intervention and those receiving the comparison intervention differently. The term "double blinding" is used to describe the masking of both subject and research personnel.

By the mid-1950s, then, through a rather circuitous and indirect route, the three elements of the randomized clinical trial were identified:

1. A comparison of individuals who receive an intervention to those who do not to determine whether the intervention makes a difference
2. The random assignment of individuals to an active treatment or comparison (often a placebo or inert treatment) group to equalize the effects of other variables that might influence outcome and remove the possibility that assignment to the treatment or control group is influenced by a characteristic that would affect outcome
3. Unawareness in both researchers and subjects of which participants receive the active treatment and which are in the comparison condition, thereby exposing the two groups to the same amount of Hawthorne effect and removing the possibility that the two groups would be treated differently

These criteria, little more than sixty years old, are used throughout science to determine the probability, compared to chance, that an observed outcome can be attributed to the experimental condition or intervention. Together, they provide a direct method of demonstrating cause and effect and quantifying the strength or likelihood of the causal relationship. However, they do not provide absolute assurance that the intervention caused the observed difference because the possibility that the difference occurred by chance or because of an unrecognized difference between the exposed and unexposed groups remains. Said more technically, the probability that any observed difference can be attributed to the intervention or experimental condition is always less than 1 (or a likelihood of less than 100 percent). As a result, Hume's claim that inferential reasoning is always used in establishing causal relationships is not refuted by the RCT. Rather, the RCT provides a method for determining the strength or confidence with which a causal claim can be inferred. The likelihood that a causal relationship exists can be further increased by additional steps, for example, repeating the study in an independent setting and replicating the results.

Today, most agencies that have the responsibility of approving drug therapies throughout the world have a requirement that a minimum of two independent studies demonstrate the efficacy of a treatment in order for it to be approved. However, there is no magic number of studies that absolutely proves efficacy, since there are many instances in which several studies demonstrate the efficacy of an intervention and several do not.

Another challenge has become clear in the past several decades. How can a finding that there is *no* effect or difference allow the claim that there is *not* a causal relationship? This question is important because it addresses the insights of Bacon and Popper that the ability to eliminate possible causal relationships strengthens the conclusion that a causal relationship exists when a difference *is* found. This issue is addressed by the concept of *statistical power*, a measure of the confidence with which it can be concluded that enough subjects were studied to detect a difference if it existed. This provides a statistical method for quantifying the strength of the inference that no causal relationship exists if no statistically meaningful difference is detected, but, again, it cannot eliminate the possibility that a negative study is "falsely negative."

COUNTERFACTUAL REASONING AND STATISTICAL METHODS FOR ORDERING SEQUENCE

Two final approaches used by epidemiologists and social scientists to identify causal relationships seek to apply the logic of the randomized clinical trial to settings in which true randomization is not possible. In the counterfactual approach, the identification of a group *not* exposed to a potential causal variable is used as the comparison (or "control") group. The real strength of the RCT method, as first recognized by Fisher, is that randomization of enough subjects to the treatment and control groups should make them equivalent in every way except the variable of interest—the treatment. The counterfactual approach makes the same assumption; if two groups being compared are the same in every way except for the exposure to the possible or hypothesized causal variable, then a causal hypothesis can be tested to explain different outcomes.

Keep in mind that randomization in RCTs sometimes "fails" because the groups being compared differ in some possibly meaningful way. For example, they may differ in the ratio of females to males. Because this is undesirable, researchers can take steps to increase the likelihood that the groups will be very similar—in this example, designing the study so that subjects are assigned to one of the groups based on whether a male or female is needed at that point to equalize the female to male ratio—but chance is always operating, and at the end of a study the groups being compared are often found to differ in some way that might or might not be relevant to the outcome. Statistical methods are sometimes used to "adjust" for these undesirable differences after the fact, but this introduces statistical manipulations that are not desirable in an ideal study. The counterfactual method faces the same challenge, and to a greater extent. It is very difficult if not impossible to find comparison groups that are the same in every way except for the causal variable of interest.

Methods and statistical techniques have been introduced in recent years to minimize and adjust for the differences between a group exposed to the variable of interest and those not exposed, but these approaches still require the researchers to identify those variables they believe, based on their expert knowledge, might be influential and compare individuals who are similar on all of these. Some such techniques also introduce statistical manipulations that change what was actually found.

Statistical techniques have also been developed to operationalize the sequential relationship between cause and effect. That is, if there are data available that have followed individuals or groups over time, and if it can be shown that potential causative variable A almost always precedes variable B and that B is unlikely unless A is present, then there is statistical support for A causing B. This is an application of the first two features of Koch's postulates and helps in situations in which an experiment could never be carried out. However, there is always the possibility that some other variable preceded A and was both necessary and sufficient to cause B, or that A led to another variable C that is the actual causative variable.

As should be clear by this point, every approach to causality has its limitations. The strength of the counterfactual approach and the statistical methods that account for chance associations, identify sequential relationships, and eliminate variables that are not of interest but are known to influence outcome are that they provide an avenue for studying potential causal variables in situations in which randomization is not possible. Their great limitations are that the groups being compared are likely to differ in meaningful but unanticipated ways and that other variables not known to the researchers could influence the outcomes. A new field, comparative effective research, has begun to apply these methods to study the efficacy of treatments that are widely used in hopes of finding whether one is more effective than another at the population level, but such comparisons rarely identify why one group received a treatment or had an exposure to a potential causal variable and the comparison group did not. Like most of the methods to assess causal relationships discussed in this book, the counterfactual approach has its strengths and limitations, and these need to be considered when the evidence that they produce is being considered.

The methodological advances of epidemiology and clinical trials thus do not eliminate Hume's insight that causal reasoning requires an inferential leap because there is always the possibility that some other variable or set of variables is the actual causal agent. Nevertheless, as Bradford Hill pointed out, the likelihood of an incorrect inference when using empirical logic is lessened when multiple approaches are combined. Several steps that can be taken to increase accuracy are the testing of hypotheses that have been stated before the experiment is carried out, the replicating of studies in which there have been positive findings, avoiding the false rejection of a hypothesis by having enough people in the groups being compared, the development of multiple lines of evidence (convergent validity), and the demonstration of the presence of a graded relationship (for example, the more A there is, the more C that results; or, the more A is diminished, the less C that results). The absolute demonstration of causality remains beyond reach, but the paralysis that Hume's claim might induce is countered by approaches that quantify the likelihood and strength of proposed causes and meaningfully increase or exclude their causal role.

ACCIDENTS

> The better and the safer technology becomes, the more we presume human error when something goes wrong. If it is not the error of the captain or crew, it is one of the engineers or designers of equipment, or of executives and their maintenance policies.
>   —Edward Tenner

The term "accident" implies that the cause of an event is out of the ordinary and that the outcome was unintended. Some scholars disparage the use of the word "accident" because it implies that the occurrence was beyond human control, but, whatever word is used, they provide another set of noteworthy occurrences in which the identification of causal agents and causal chains is a primary goal. Two books highlight the challenges faced by those studying accidents.

In *Normal Accidents*, the sociologist Charles Perrow focuses on events that cause relatively substantial damage (he refers to events that have minor adverse outcomes as "incidents"). He divides them into those with straightforward causes, such as single actions or failures of single parts, and those with more complex causes, for example, those that result from the interaction of several elements of a system. Perrow provides an in-depth analysis of incidents and accidents in nuclear power facilities and identifies four types of causes. First-level incidents or accidents result from the failure of a *single part*, such as a valve. Second-level accidents result from the failure of a collection of related parts that make up a *unit*. The failure of a steam generator is an example. Third-level accidents involve the failure of complex *subsystems*, for example, the cooling system of a power plant. Fourth-level accidents arise from design errors that affect whole systems.

Several points of Perrow's analysis are instructive. One is that its levels are somewhat parallel to those proposed by Aristotle. The failure of a valve or steam generator (a first- or second-level accident) can be a provoking cause because its failure leads directly to the incident, while the failure of a unit or subsystem could be either a predisposing or precipitating cause depending on whether a design flaw leading to system failure existed before the accident occurred or was itself the result of the interaction of several elements of the system. Perrow even invokes purposive cause in his fourth level because he concludes that humans build complex systems in spite of the knowledge that their failure could result in widespread, devastating damage—that is, he invokes hubris, greed, or human characteristics or behaviors as underlying the push to build bigger, more complex structures and systems even though they have a meaningful chance of failing at some time and, in some circumstances, leading to significant harm.

Perrow's schema makes the point that the examination of accidents at several levels of analysis aids in both the analysis of causality and in communicating the findings of the analysis. Because the primary purpose of his schema is to provide a general framework for analyzing accidents, the ultimate criterion for judging it is whether it is helpful, not whether it represents something that exists in nature. The same can be said for the Aristotelian schema: utility is the ultimate standard by which it should be judged. For Galileo, the Aristotelian model was an impediment because it led people to focus on questions such as ultimate cause that could not be assessed experimentally. History suggests that Galileo was right at the time, because science flourished in the years after his pronouncement in ways it had not before (a causal attribution based on narrative reasoning). Today, methods are available for analyzing complex interactions at a systems level and make the Aristotelian model more useful because it provides a structure for analyzing complex issues. Multilevel analysis is still more challenging than level 1 analysis, but it is a more powerful tool of analyzing the many complex situations we are interested in.

Perrow's analysis echoes another point that arose earlier: the level at which an analysis should be carried out depends, in part, on the issue being addressed. Thus, the analysis of a single accident might focus on specific aspects of that power plant or of that failure, but an analysis of multiple failures or of the failure of a large power grid might need to focus on systemic or programmatic interactions and analyze issues as far ranging as patterns of electricity use by the populace, climate, governmental policy, and the availability of materials for manufacture. However, it should also identify the precipitating cause that initiated the failure if possible.

The analysis of complex accidents also highlights some of the issues that emerged in the discussion of complexity and nonlinearity. Some parts, units, and subsystems serve multiple functions and interact with several or many other parts, units, and subsystems via feedback loops and branching paths. Even though these may involve only a small portion of the system, their interactions can result in the rapid, exponential magnification of an untoward event. That is, accidents may reveal a nonlinear relationship that exists between a single event and its outcome. The greater the number of such interactions within a system and the larger the system, the higher the likelihood that nonlinear causal dynamics will be operative.

For example, if failures occur in two elements, problems can arise that were never anticipated, a situation Perrow calls "interactive complexity." If, in addition, the elements of the system are "tightly coupled," the error is likely to spread so rapidly that it cannot be contained. Perrow believes such outcomes are inherent and unavoidable in complex systems—hence his use of the term "normal accidents." He notes, though, that isolated problems can arise in a complex system, and thus the use of linear, categorical logics and methods might be appropriate even in very complex systems. If this approach is successful, it is desirable because of its simplicity and limited use of resources. The ultimate elimination of smallpox, for example, required developing the means of identifying the few remaining pockets of infected humans, even though the causal web included aspects of the virus's unique biology, human immunity, patterns of human travel, and the reluctance of governments to allow outsiders on their soil.

Perrow concludes that accidents cannot be totally eliminated because it is not possible to anticipate all interactions, but he does believe that their frequency and severity can be lessened by careful design and vigilance.

Edward Tenner suggests a broader approach to the study of accidents, defining them as events that have unintended consequences. In his view, these unintended consequences can be either desirable or problematic, a counterpoint to Perrow's pessimism about the inevitability of failure. Tenner points out, for example, that the slowing of highway traffic by congestion has been paralleled by a decline in the rate of fatal accidents per mile driven throughout the twentieth century. The rising frequency of automobile accidents as people drove more miles was a spur to the development of safer cars, roads, and driving practices, and the increasing complexity of automobiles and the highway systems needed to support them has resulted in a continued decline in accidents per mile driven. Although such "progress" is not inevitable (the drop in accident rates in the airplane and automobile industries has not been matched by declines in accident rates in the marine industry, for example), Tenner concludes optimistically that adverse outcomes seen as disasters often spur changes in rules and technology that result in improved safety and health. He concludes that "when it comes to interpreting the last hundred years, the optimists have the upper hand. . . . Pessimists welcome emergency as a violent cure for profligacy. Optimists welcome it as an injection of innovatory stimulus."

9

NARRATIVE TRUTH: THE EMPATHIC METHOD

The study of history is a study of causes.

—Edward Hallett Carr

There can be no definitive history anymore . . . [but] just because we all have a different idea of what history is, or should be about, does not mean that we no longer read one another's works.

—Richard J. Evans

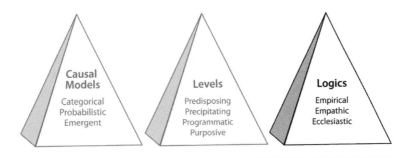

Like so many things, the recognition that there are major differences between history and science can be traced back to Aristotle. However, in the modern era it was the early eighteenth-century philosopher Giambatista Vico who first emphasized the distinction. The differences were further contrasted and highlighted by the nineteenth-century sociologist Max Weber, and, today, the distinction is often depicted as absolute.

The nonscience form of knowing is referred to by many different names—history, narrative, story, chronicle, empathic method—and its essential methods are used in many disciplines. While there are subtleties of meaning that distinguish one from the other, this chapter will use these labels interchangeably, because the goal of the chapter is to identify what these disciplines and techniques share in common, particularly in relationship to their approach to causality.

Unfortunately, the search for a succinct definition of narrative/ historical reasoning will ultimately fail, just as the attempt to define science succinctly and accurately did earlier. But just as the first chapter was able to identify salient elements of the scientific method, so too will this chapter identify the central features of the narrative method of knowing. This approach of painting broad outlines but not filling in the details of the definition will dissatisfy some (to the point that some will continue to believe that there is no distinction between scientific and narrative knowledge, while others will consider the distinction absolute and the failure to specify the differences a failure of effort), but it reflects one of the central themes and biases of this book: making specific and absolute distinctions between or among ideas is not a necessary requirement for demonstrating meaningful distinctions. In fact, this approach mirrors a major assumption of the approach to causality proposed in this book: the greater the evidence that a distinction exists, the greater the likelihood that the differences are meaningful, even "real." The philosopher Stephen Toulmin (1922–1999) suggested that claims such as these be referred to as "practical arguments" as opposed to absolute arguments, and the claim I am making here is made in that spirit.

Δ Δ Δ

The founding of history is often traced to the fifth-century BCE Greek writer Herodotus, and, until the last century or so, most historians functioned with the certainty that their summaries of and conclusions about past events were true and accurate. However, the realization among both historians and nonhistorians at the end of the nineteenth century that reconstructions of the past, even by professionals, are significantly influenced and shaped both by the belief system of the individual doing the reconstruction and by the attitudes of the era or setting in which the person or group is working became more widely accepted. This "discovery" paralleled the recognition among physicists at the beginning of the twentieth century that absolute certainty is not possible in their discipline. However, the reaction among practitioners of these two disciplines seems quite different. Some professional historians and some critics of their way of knowing and determining causality attacked the very usefulness of the discipline and of its methods rather than following the path of Weber and twentieth-century theoretical physicists in accepting these limitations as inherent to the subject matter. One can gain a sense of the denigration of this type of knowledge by considering the negative sentiment that usually underlies the adjective "rhetorical" and the phrase "Whig history" (a reference to the surety with which conservative British historians and politicians of the nineteenth century expressed their views and the later rejection of these views by individuals with other views), both of which have been used to describe the relativity of the historical method.

The level of this skepticism is captured in this tongue-in-cheek quote from the book *What Is History?* (1961), the compilation of a lecture series given at Cambridge University by the eminent British historian Edward Hallet Carr (1892–1982): "I hope that I am sufficiently up-to-date to recognize that anything written in the 1890s must be nonsense. But I am not yet advanced enough to be committed to the view that anything written in the 1950s necessarily makes sense." The sentiment expressed by Carr sarcastically captures the dilemma that can result when absolute accuracy is used as the only standard for judging the value of a discipline and its methods—if absolute accuracy is impossible, then accuracy does not exist. Of course, even those who believe that science is moving toward an ever more accurate description of how nature works would acknowledge that absolute knowledge has not yet been attained in any field of science either, and this raises the possibility that it may never be.

The alternative view, the one espoused here, is that absolute certainty is unattainable via either the scientific or the empathic/narrative method. What sets science and history apart is the subject matter they study, many (but not all) of the methods they use, and the end products they seek. This chapter will seek to demonstrate another similarity: knowledge in history, as in science, can become increasingly accurate over time and, more importantly, the likelihood of accuracy can be increased by the use of appropriate methods and safeguards. In chapter 10, I will argue that the inability of either the narrative or the scientific method to attain absolute knowledge or identify final causality is the major factor that distinguishes them from the religious or ecclesiastic method of knowing.

Not surprisingly, Carr tries to answer the question that makes up the title of his book. He begins by noting that historians rely on facts but goes on to claim that the collection of facts is not their main function. Rather, the primary attribute of well-respected historians is how well they put the facts together and how they interpret them. Factual knowledge can change over time, he emphasizes, as new information is unearthed (new documents or witnesses can be found, for example). This new information might or might not change the interpretation of the facts that were already known. The central and distinguishing act of the historian, however, is the linking and interpreting of the known facts.

At one level this seems similar to science, since science both seeks to determine "truth" through the accumulation of facts about nature and assumes that truth becomes more accurately represented as more facts are compiled. Newton's great insights on gravity, for example, while still taught to beginning students, have been superseded by modern ideas that depend both on fields of mathematics developed three hundred years after he died and by discoveries in the discipline of particle physics that emerged in the twentieth century. This hasn't diminished the reverence or appreciation with which Newton's ideas are held. However, since there is no comparable person to Newton (or Galileo or Einstein) in the pantheon of historians, there must be other differences between narrative knowledge and scientific knowledge that make the distinction useful or important.

One clear difference, as illustrated by the work of Newton, Einstein, and Galileo, is the distinction that Carr made: the discovery of facts has a central, essential role in science, and the discoverer of important facts can gain renown. In contrast, fact discovery generally plays a secondary or minor role in history and other narrative disciplines. Indeed, it might be said that the ultimate goal of science is to discover the facts that describe all of nature, whereas the ultimate goal of history is to gain as comprehensive an understanding as possible of how past events led to later occurrences. While historians have tried to find a "unified field theory of history" from time to time, the idea that there is a single pattern of history that is waiting to be discovered is now rejected by most professional historians and social scientists, whereas this remains a goal of those theoretical physicists who are exploring approaches such as string or brane theory, or a "theory of everything."

Since the narrative disciplines do rely on the accumulation of facts, the replication of or agreement about the veracity of these facts is an important aspect of the historical method. Likewise, skilled sociologists, political scientists, and economists gather facts about their fields of interest and weave them together into coherent chronicles or stories. However, different observers can examine and describe the same document, event, or set of facts and interpret them quite differently, even when they agree on the meaning of the words. There may be no resolution of the disagreement, or, if there is one, it comes about when a consensus of colleagues favors one interpretation over another. In science, in contrast, the ideal way to resolve a difference of opinion or interpretation is to design and carry out a new experiment that excludes some possible explanations and verifies the predictions of others. In the narrative disciplines, new facts settle disagreements only infrequently, and neither replication nor prediction is essential to the success of a hypothesis; in science, the results of a new experiment or the confirmation of a prediction can overturn even the most widely accepted scientific idea. Thus, both approaches depend upon the integration of multiple facts but, as Popper emphasized and Bacon mentioned four centuries before, predictive falsifiability is central to many disciplines that utilize the scientific method but not those that depend upon the empathic approach. As Carr emphasizes, the skilled historian is one who draws from the facts a convincing interpretation that increases the understanding of a broader question.

A rather trivial observation highlights this distinction. In science, the discoverer of a fact or the designer of a single experiment is awarded the Nobel Prize if the results are judged to be novel enough, even if others develop the finding into a broader understanding of the structure of nature. In history, the discovery of a new document or "fact" can be important or exciting, but it is the person who interprets its relationship to other documents or information and links them together who is acknowledged as the expert. Carr describes the role of fact accumulation in history as a tension between the "Scylla of an untenable theory of history as an objection of facts . . . and the Charybdis of an equally untenable theory of history as the subjective product of the mind of the historian." His conclusion is that "history is a continuous process of interaction between the historian and his facts."

In the end, the roles of the discoverer and of the facts are mirror opposites in the two approaches. In the narrative disciplines, the individual telling the narrative or story remains a central and necessary participant; in the empirical sciences, the role of the individual discoverer becomes less important than the goal of obtaining as complete an inventory of facts as possible. Even though individuals are central to the discovery process in the empirical disciplines, any discovery could be made by someone else. Not so in the narrative disciplines. Scientists seek a representation of the natural order that exists whether or not they are studying it. Historians seek a comprehensive understanding of events by dint of the individual's ability to link together convincingly what is known. Carr concludes that good historians acknowledge that the current worldview influences their interpretations and take this into consideration in presenting their synthesis. Scientists, I suggest, operate on the assumption that their conclusions represent an approximation of truth that is limited by a lack of knowledge of all potential facts. The empathic, narrative approach will always be strongly influenced by individual practitioners and the cultures in which they work. The empirical approaches seek to minimize these as much as possible. (The discipline of the history of science often combines the two approaches, however.)

If history is more (but not totally) "subjective" and science more (but not totally) "objective," is one inherently better or more useful than the other? Both Vico and Weber thought not, and I hope the previous paragraphs make the case that they are right. The two ways of knowing address very different issues and thus must rely on different approaches. When dealing with individual human beings in specific circumstances or with causal questions about human motivations, the empathic approach is more powerful. When dealing with the nonanimate world or with human behaviors at the level of probabilistic prediction, the empiric method is often more useful. When dealing with the interface, for example, between molecules and choice, both approaches have their strengths and limitations.

In a 1997 book aggressively titled *In Defense of History*, the British historian Richard Evans explores the methods of his chosen discipline and also identifies similarities between scientists and historians. Usefully, he emphasizes other differences particularly relevant to understanding the role of causality in the two approaches. First, Evans notes that history is "a bad predictor of future events . . . because history never repeats itself." In this he disagrees with Carr and the historian George Santayana. Second, "history cannot create laws with predictive power . . . [for while] history . . . can produce generalizations . . . the broader they are, the more exceptions there are likely to be and the farther removed they will become from hard evidence which can be cited in their support." This is not a negative for Evans, though, since he concludes with the positive assertion that "life, unlike science, is simply too full of surprises."

Even on these points, though, universal agreement among historians is lacking. For example, John Lewis Gaddis, a distinguished American historian, proposes a middle ground, citing Machiavelli as the source of the idea that "history is arguably the best method of enlarging experience." He thus proposes that "studying the past is no sure guide to predicting the future. What it does do, though, is to prepare you for the future by expanding experience, so that you can increase your skills, your stamina—and, if all goes well, your wisdom."

THE NARRATIVE METHOD

If the general outlines of the story are sound, that is all that our present purposes demand.

—Stephen Toulmin

The primary tool of the historian is the narrative method, defined by Richard Evans as the combination of "action, happening, character and setting." A similar definition has been proposed by Paul McHugh and Phillip Slavney, two of my teachers, in their book *The*

Perspectives of Psychiatry. To them, the central characteristic of the narrative or life-story method is the linking of setting, sequence, and outcome into a coherent, comprehensive web.

The skill that sets apart the average from the excellent user of the narrative method, as Carr noted, is the ability to make linkages that convince other experts of the power of those linkages. This use of rhetoric has been recognized as a skill at least since the ancient Greeks, but, as noted at the beginning of this chapter, today the word "rhetoric" is often used in a negative context to indicate that a statement is "mere words" and that any conviction that arises from such a discussion is tainted by undue influence as opposed to the sagacity and beauty of the argument. Implicit in this denigration of rhetoric is the idea that beliefs based on "facts" are somehow different, an opinion that seems to be an extension of the idea that scientific knowledge is better than or more objective than knowledge gained through discussion and argument.

Of course, scientists must convince others of the validity of their specific conclusions, and they use the tools of rhetoric to do so. As already noted, science has the techniques of predictive verification and falsification to bolster the validity of an argument, evidence not available in the narrative disciplines because the events that are studied are often unique or different enough that they cannot be considered comparable or identical. Nonetheless, the role played by rhetoric in scientific debates about the validity of a line of thinking is often underappreciated and even ignored.

Conversely, the accuracy of conclusions derived using narrative methods can often be confirmed or refuted by relying on relevant documents or testimony from multiple sources, by finding similar patterns of outcomes in different circumstances, or by showing that the differences between or among a set of events being compared are inconsequential or minor. This bears some resemblance to the epidemiologic concepts of the counterfactual, convergent validity, and preponderance of evidence discussed in chapter 8.

A major strength of the narrative method, perhaps its greatest advantage, is its ability to increase our understanding of unique past and present events. Perhaps the strongest evidence of the power of the narrative approach is its universality, that is, its use by human beings in all cultures. Science derives its power from its ability to examine events that can be repeated and manipulated and from its capacity to derive knowledge that predicts future events and leads to the development of products that benefit humankind.

Many scientific disciplines do focus on unique, past events such as the origins of the universe, but when they do, they combine historical and scientific methods of knowing. When historians collect documents that confirm or refute their hypotheses, they are engaging in a similar act. In both instances, the interpretation of the content, implications, and importance of the evidence and the acceptance of the proposed interpretation rests upon the reasoning and rhetorical skills of the practitioner. Thus, narrative and scientific disciplines share reliance on accurate facts and the use of rhetorical methods to convince others of the accuracy of proposed linkages.

A shared reliance on rhetoric is not the only similarity between the two approaches. The absolute dichotomization of disciplines into empiric or empathic is also undermined by the existence of many disciplines that straddle the divide between science and the humanities. In paleontology, for example, the discovery of new fossils leads to new interpretations and, sometimes, to new proposals about past species modifications. These interpretations are neither verifiable nor refutable, at least no more than are historians' reconstructions and interpretations of past events. New fossils, like new documents, are interpreted in light of those discovered previously. Future discoveries will likely modify the interpretation again. Evolutionary psychology is another field in which post hoc explanations rather than predictions are the primary focus of the intellectual work. Even economics, a discipline that has a primary focus on making predictions, primarily uses past economic patterns and the tendencies of humans to behave in certain ways as the primary basis upon which predictions about future economic activity rest. It identifies relationships among variables, not mechanisms that are inherent in nature. Some will disagree with this opinion, but the predictions of economists seem no more accurate in the long run than those made by political scientists, historians, or sociologists. Each of these disciplines straddles the empiric and the empathic, relying upon the strengths of each to advance knowledge and understanding and make more accurate predictions.

The distinction between the narrative and empirical approaches is even more blurred because much of science does not, and sometimes cannot, rely on replication or falsifiability. Heisenberg's uncertainty principle is an example of a powerful scientific idea that does not lend itself to refutation or positive proof. Likewise, for most of the first hundred years after Darwin and Wallace proposed the theory of natural selection, the study of evolution was considered a scientific discipline on the basis of its comprehensiveness and the lack of competing alternatives rather than because of the positive results of experiments or refutation of alternative ideas. Its acceptance did depend upon the weaving together of different facts and multiple lines of evidence, that is, convergent validity, but the mechanisms by which evolution operated within populations are just now being explicated. (The idea that evolution is a powerful empirical and empathic idea will be developed further in chapter 11.)

Another example of the blurred distinction between the empathic and empirical methods of knowing is the use by scientists of terms such as "beautiful" and "comprehensive" and the invocation of such adjectives in justifying their approach to acceptance of a theory. Albert Einstein's thirty-year search for a unified field theory that would explain the four basic forces of nature (strong, weak, electromagnetic, and gravitational) reflects a belief shared by many scientists that "simplicity is beauty," an aphorism sometimes referred to as the "law of parsimony" or Occam's Razor, after the fourteenth-century writer who espoused it. As Brian Greene ably depicts in his book *The Elegant Universe*, string theorists and brane theorists persist in the same quest. This speaks to the power of the human tendency to define some activities as inherently "more appealing," "right," or "better," even in the hard physical sciences, a subjective, nontestable idea and rhetorical construct if ever there was one.

THEN WHY NOT ABANDON THE DISTINCTION?

While the empiric and empathic approaches share many elements, the claim that the boundary is so fuzzy that the distinction is of no value is an error. Acceptance of a narrative idea rests primarily on how the community of scholars or experts assesses the reasoning that produced and supports it. Facts can be verified, alternative explanations can be sought, and the strength of the links examined, but the inability to test or refute the claimed association (as opposed to the facts, which can be confirmed) does distinguish the narrative, empathic disciplines from empirically verifiable science. Conversely, there are topics such as the behavior of individuals and groups to which the narrative approach can contribute an understanding and that the empirical sciences have thus far failed to explain with an accuracy that would meet the standards of science. For me, though, the most convincing evidence that the distinction has value and says something important about the structure of knowledge is the fact that the narrative approach is present in all cultures and used by all individuals, whereas the methods of science are a relatively recent invention.

Studies carried out by the neuroscientists Roger Sperry and Michael Gazzaniga offer further support for this claim. They examined patients who had previously undergone "split brain" surgery, that is, had the large fiber bundle that connects the two hemispheres or sides of their brain severed in an attempt to stop the spread of seizure discharges from one side of the brain to the other; they found evidence that there is a "center" in the brain, near or overlapping with the language area in the left hemisphere, that naturally "makes" connections between disparate pieces of evidence. This strongly suggests that the human brain is constructed to carry out narrative reasoning and that the linking together of facts into a narrative causal web is innate, just as Kant suggested.

The scientific method, in contrast, seems to be a new construct that did not exist before the Enlightenment, even if historical threads can be traced back thousands of years. It is a learned method that has evolved and been refined over many generations. Our understanding of the scientific method is quite different than it was fifty or one hundred years ago—the uncertainty principle and the logic of the randomized controlled trial are two examples discussed in this book—and it will continue to change as new analytic tools and perhaps conceptual advances are developed. Perhaps narrative reasoning can be said to be "hard wired" while scientific reasoning is a learned approach. Both have been improved by the accumulation of knowledge that occurs when communities of scholars cooperate, argue, invent, merge, and disentangle the useful and accurate from the unreliable and unpredictable. It is their shared and disparate underpinnings that convince me that keeping the two constructs separate remains not only useful but necessary.

Recognizing that each approach has different strengths and limitations also counters the claim that evidence gathered by one type of reasoning is somehow better or more powerful in the aggregate than the other. The outright dismissal of one method or the claim of ultimate superiority of one over the other ignores the power of each approach to identify causal relationships in different circumstances.

For example, a claim could even be made that causality itself is the search for a coherent narrative and therefore an empathic concept. However, this ignores the unique ability of the scientific method both to exclude some proposed causal mechanisms and to assign mathematical likelihoods to other hypothesized causal relationships. Nonetheless, the construct of causality requires an acceptance and embrace of the idea that events are linked and that the sequence of events is responsible for the outcome because this claim cannot be posed as a scientifically testable idea. This is the reason why it was necessary in chapter 1 to posit or assume the existence of causality. Needing to begin a discussion about causality with a nonprovable idea as a starting point only demonstrates that ultimate knowledge cannot be a goal in the search for causal understandings of the world in which we live.

In summary, the distinction between empirical, scientific disciplines and empathic, narrative disciplines is not absolute. Each method is best suited to examining certain kinds of questions and each is open to abuse (scientists believing they should have unique input on political questions because their methods are objective, or social scientists believing they have unique input to political decisions because they have studied past events). Emphasizing the misuse of either approach as a means of denigrating its value is ultimately foolish. In the end (or, better yet, at the beginning), each should be appreciated for its strengths. The very notion of causality bridges the empathic and empiric approaches to knowledge and uses them as tools to discover causal linkages. The next three sections illustrate uses of the narrative method and how facts can be verified, how some narratives can be valued more highly than others, and how the changing interpretations over time of one work of narrative strengthens rather than denigrates its value.

HOLOCAUST DENIAL AND THE TRUTH OF HISTORY

A libel case heard in an English court in the year 2000 provides an opportunity to scrutinize the methods of the historian and the limitations and strengths of the narrative methods. This discussion (it is not a history, as I have not performed original or extensive research on the topic) is based on the book *Lying About Hitler: History, Holocaust, and the David Irving Trial*, by the British historian Richard Evans, whose views on the concept of truth in history have already been mentioned. Evans himself was a participant in the events, as he was the chief defense witness, so he cannot be said to have come to the subject from a neutral point of view. And of course, no single event should be used to defend and define such a broad concept as historical truth. Nonetheless, the involvement of the court provides some semblance of neutrality, and the strict libel laws in England (it is up to the defendant to prove that libel did not occur, a requirement that favors the plaintiff—though this may be revised in the future) make this an interesting event to examine.

The case involved the American historian Deborah Lipstadt, whose book *Denying the Holocaust* examines the movement denying that the Nazis had systematic plans to wipe out groups of individuals such as Jews, Gypsies, and other "inferiors" that it considered enemies of the state. In the book, Lipstadt makes the claim that David Irving, a prolific but not traditionally trained author of historical works about Germany in World War II, falsified documents to bolster his claim that the Holocaust did not happen. When the book was published in England, Irving brought suit claiming libel.

Lipstadt's counsel based its defense upon the claims that Irving had persistently and purposefully falsified his translations and that he had made other untrue statements about documentary evidence. (They could alternatively have offered the defense that her statements were being misinterpreted and were not a criticism or that they were not harmful.) Irving claimed, for example, that Hitler did not know of the planned exterminations until 1943, long after a strategy to implement them had been made and launched. Using numerous retranslations of the documents that Irving had relied upon, and showing that Irving had altered the well-established and widely accepted sequence in which the events under discussion had occurred, Evans claimed in court that Irving had purposefully altered evidence to support his claims about the lack of Hitler's culpability in the planning and execution of systematic exterminations. In the end, the judge based his opinion on the falsity of the evidence that Irving had produced and the truth of Evans's claims, and he decided for Lipstadt and against Irving.

Several aspects of this example help explicate the criteria for accepting causal explanations when using a narrative approach. First, documentary evidence can be viewed by multiple individuals and a consensus developed about the content of the material. Second, experts and knowledgeable individuals can and should be expected to support their claims with evidence that can be verified, and multiple instances of an event suggest a pattern rather than a random error or event. Evans was able to convince the judge that Irving had mistranslated many documents and that these incorrect translations were always in the direction of supporting his claim; they were not random errors. In doing so, Evans was able to develop a convincing case that Lipstadt was correct in her claim that Irving had falsified documentary evidence. Third, the case demonstrates that the scrutiny of evidence by other experts can serve as a "self-correcting" mechanism in historical research, just as nonreplicability does in science. Finally, it demonstrates that egregious falsification can be identified in narrative disciplines just as it can be in science and negates the claim that "anything goes" in narrative reasoning.

This case also highlights differences between the empirical/scientific method and the empathic/historical/narrative method. Convincing one judge of the accuracy of Lipstadt's claim is far different than the widespread acceptance of a causal scientific theory based on replicable experiments and observations, elimination of alternative explanations, convergence of multiple lines of evidence, and agreement among multiple experts. Second, historical documentary evidence can be forged, altered, or purposefully produced to mislead, and this can be difficult, perhaps even impossible, in some cases to demonstrate. Of course, laboratory notebooks and photographs of experimental results can also be purposefully constructed to mislead or lie, but if the experiments or measurements can be repeated, then an avenue to test a claim is available that is not available to the narrative historian because the latter is dealing with events that occurred in the distant past and are unique, that is, occurred once and are nonreplicable.

THE WRIGHT BROTHERS AND THE INVENTION OF THE AIRPLANE

The year 2003 marked the one hundredth anniversary of the invention of the airplane. Predictably, a number of books were published to mark the centenary and sought to address an intriguing question, which was stated clearly in the first two sentences of one of the books, *The Wright Brothers and the Invention of the Aerial Age* by Tom D. Couch and Peter L. Jakab: "Why Wilbur and Orville? How did these two modest small businessmen, working essentially alone, with little formal scientific or technical training, solve a complex and demanding problem that had defied better-known experimenters for centuries?"

Couch and Jakab's book offers a sophisticated and wide-ranging answer to this question by identifying multiple factors that contributed to the Wrights' success. The authors cite their mother's mechanical aptitude as both a genetic and an environmental influence. They also cite the brothers' upbringing in a loving, supportive family that valued hard work, individuality, and independence as fostering a devotion to each other and their dogged quest to build a flying machine.

Couch and Jakab make the claim (convincingly, I suggest) that the brothers' intimate knowledge of the bicycle provided them with their crucial insight that when bicycle riders turn, they do so in multiple planes, simultaneously leaning into the turn and turning the wheel in the direction of the turn, in contrast to buggies, which are turned in a single, upright plane. This observation led them to analyze bird flight carefully and discover that bird wings deform when they turn. Combining these observations, the brothers concluded that the airplane would need to operate in three planes and that the pilot would also have to have controls that allowed the wings to be moved vertically, horizontally, and laterally.

Couch and Jakab also cite the brothers' practical experience in having established a print shop, designing and building their own printing press, and working on bicycles as factors in their commitment to trial-and-error engineering and their attitude that unsuccessful approaches should be abandoned and successful ones sequentially added together.

The brothers' final success thus combined their knowledge of bicycle handling, their careful observation of nature, and the mechanical skills they had developed into an engineering design that launched a revolution. This attitude also applied to the trial-and-error learning that they applied to developing the skills necessary to fly the airplane they built.

Several of the other books published at the same time discuss some of these issues. What makes the Couch and Jakab book so successful is its weaving together of the authors' knowledge of aeronautical engineering principles and their depiction of the Wright brothers' upbringing, personality characteristics (and those of some of their competitors), work ethic, work experience, and observational skills into a coherent, convincing, and engaging narrative. Are the authors correct in their analysis? To this nonexpert reader, the answer is a resounding yes. Might other information be discovered in the future that could change their analysis? It is certainly possible because their history can be no more complete than any branch of science. But like areas of science that no longer hold much interest after the main questions have been answered, their biography is so satisfying and comprehensive that I think it unlikely that the answer to their initial question "Why the Wright Brothers, and why that point in time?" will be answered in a radically different manner in the future. Couch and Jakab raise a causal question, and their book provides a sophisticated and satisfying, albeit not final, answer.

MANY DE TOCQUEVILLES

Alexis de Tocqueville's (1805–1859) *Democracy in America* is a two-volume work that describes the author's visit to the United States in 1831 and his conclusions about the strengths and weakness of the democracy that he observed. It remains a widely cited work 180 years after its publication. Among the highlights of de Tocqueville's analyses is his being the first to note that democracy's emphasis on individual initiative could lead either to great individual success, a major strength of American democracy in his opinion, or to social isolation, a significant downside to that form of governance. He also suggested that democracies had a tendency to develop a tyranny of the majority but felt that this drawback was balanced by a strong bureaucracy, a central feature of democracies in his view.

In an article written about the book in 1976, the Columbia University professor of humanities Robert Nisbet (1913–1996) noted that there were actually *"many Tocquevilles."* Reflecting a point made earlier in this chapter, Nisbet notes that de Tocqueville's two volumes have been interpreted very differently by different individuals in the decades since its publication. They were initially received quite enthusiastically and widely admired for about twenty-five years, but "from the late 1860s until the late 1930s only an occasional monograph or article on Tocqueville appeared. But . . . around 1940 . . . came the deluge. . . . By the late 1940s it was a rare month that did not yield treatments of, or references to, Tocqueville." Most notable, though, was that the focus of those citing the work changed considerably. In the 1930s, many writers cited de Tocqueville's writing on the "the masses as the source of despotism" in discussions of the rise of fascism, while after World War II writers focused on de Tocqueville's discussion of the affluence of the middle class in a democracy. In the 1950s, the emphasis changed again, this time focusing on his statements about social and cultural isolation in the American democracy.

Nisbet does not say that the writers of any one period totally ignored or exclusively emphasized a single aspect of de Tocqueville's thought. Rather, he makes the point that the emphasis of scholarship changed as the dominant issues of the culture changed. Different

writers at different times emphasized different aspects of de Tocqueville's writings as evidence of his great insight. Nisbet sees this both
as evidence of Tocqueville's great strength and of the ability of writers to find in de Tocqueville's writing support for their particular
point of view, sometimes even different or opposite in conclusion to
insights emphasized by other writers at other times. Nisbet's main
point is that a single document can be used at different periods of
time and by different individual interpreters to support very different conclusions, a reflection of how the narrative method depends on
the interpreter as much as it does on the document.

More recently, in 2007, the newly elected president of France,
Nicholas Sarkozy, used de Tocqueville to support his call for insti-
tuting political change and exhorted the public to develop new
approaches to the problems that he believed he had been elected
to solve. In a front-page *New York Times* story on July 22, 2007, his
finance minister, Christine Lagarde, is quoted as "citing Alexis de
Tocqueville's 'Democracy in America' [in saying] . . . the French
should work harder, earn more and be rewarded with lower taxes if
they get rich." Her portrayal seems accurate; that is, de Tocqueville
does mention each of these as a beneficial outgrowth of American-
style democracy as he saw it 180 years ago. But Lagarde was not
quoted as mentioning the downsides that Tocqueville also cited.

The mention of an almost two-centuries-old political science text
in such a context speaks to the depth of the insights and prescience
of its author (and probably to his French background). However,
the range of interpretations over the past 180 years highlights the
importance of who is doing the interpreting, when the interpreting
is taking place, and even who is reading the interpretation—issues
common to all works of narrative. Reading Nisbet's article today
makes it clear that Nisbet was writing at the height of the Cold War
and that this led him to emphasize those issues that were most relevant to that context. An author citing de Tocqueville today would
likely emphasize still other issues. What scholars and politicians do
agree about is the power of Tocqueville's observations and the conclusions he draws from them. What people have differed on over
time is which observations they emphasize and what interpretations

they draw from them. A world without de Tocqueville's masterpiece would be a diminished place.

THE ROLE OF THE NARRATIVE IN
CAUSAL REASONING

Causal narratives seek to knit together disparate observations, facts, and events into a coherent and inclusive whole that convincingly links later events to prior events. This method appears to be universal to human beings and may well be innate, that is, dependent upon how the human brain is structured. The power of the method is illustrated by the flow of history and the linkages that human beings inevitably find between past and present.

The three examples above illustrate some of the strengths and limitations of this approach to causal understanding. Couch and Jakab's analysis demonstrates the power of the narrative method to answer questions that cannot be answered using any other approach. A comparison of their book to others published around the same time suggests that their effort was more successful in identifying causal chains and influences than others—at least in the opinion of one reader—and that narratives can be judged on their ability to convince readers and knowledgeable scholars of their comprehensiveness and accuracy. Other readers might find the conclusions of other authors more convincing, and, unlike science, in which purported causal influences can be subject to confirmation and refutation, the narrative of Couch and Jakab must stand on the strength of its rhetoric alone. In the end, any conclusions about the power and value of the book are up to the individual reader, but this neither diminishes the beauty of its arguments or, perhaps most importantly, the pleasure of contemplating its conclusions.

The Lipstadt trial and Richard Evans's account of it illustrate the important point that there are standards for accuracy in the historical methods and that these can be used to scrutinize controversial claims. Relatively neutral individuals can examine the data used to construct historical narratives and make judgments about their validity. The narrative method is not solely dependent on the opinion of a single individual, the time or context in which it is written, or the rhetorical beauty of its argument, but each of these does influence both its content and how it is interpreted.

Finally, the lasting attention and praise directed toward Alexis de Tocqueville's insights in *Democracy in America* illustrate the power of narrative to explain wide-ranging phenomena. The work is better remembered and has had more influence than most of the science published during the years in which de Tocqueville made his trip and wrote the two volumes, and its influence upon important thinkers of later eras is a commentary on the accuracy of his comments. Yet the comparison to the writings of scientists working at the same time is unfair and inappropriate. The power of the narrative chronicle is quite different from that of science; to compare it to the importance of Mendel's discovery of the regularity of inheritance seems totally wrongheaded and unfair both to Mendel and de Tocqueville. In the end, the narrative approach to causality is a useful, even powerful, tool, just as the scientific method has propelled humans to an understanding of the workings of nature beyond anything in the past. Each approach has its place. Appreciating the distinctions between the methods they use and the differences in logic that they rely on can help prevent practitioners of each approach from overstepping their claims on causal knowledge. Recognizing that they differ in how they are applied should diminish the dismissal that is sometimes directed toward them. The question to ask is whether they are applied appropriately to queries about causal relationships, not whether either approach is the best or only way to answer questions about causality.

10

CAUSE IN THE ECCLESIASTIC TRADITION

The proof of a knowable thing is made to either the senses or the intellect, but as regards the knowledge of God there can be neither a demonstration from sensory perception, since He is incorporeal, nor the intellect, since He lacks any form known to us.

—Meister Eckhart (c. 1325)

I came to spiritual belief only very much later when I was plagued by the solution to the question of purpose.

—Allan Sandage

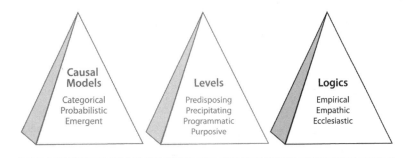

Meister Eckhart presents a challenging dilemma that is still insurmountable almost seven hundred years after he stated it—many individuals report that they gain profound emotional and intellectual benefits from their religious and spiritual beliefs, yet, he claims, these cannot be used to argue in favor of their truthfulness. Conversely, he argues, those who reject the validity or utility of religious ideas cannot use empathic or scientific/empirical arguments to support their views, either.

The essence of Eckhart's argument is that a different approach must be taken to understand causality in religion, one that is based not on logic or emotion but on an alternative type of reasoning or knowledge. The central thesis of this chapter echoes Eckhart's claim and proposes that this third approach is centrally relevant to a model of causality that has historically been present as long as recorded history and is still widely embraced throughout the world today. This chapter will argue that this alternative approach coexists with empirical and empathic approaches to causal reasoning and that it provides a third alternative logic or method of causal reasoning.

The terms "religious" and "spiritual" are sometimes used interchangeably because both refer to belief systems that seek to explain the origins of the lived world and provide guidance about how life should be lived. However, the term "religion" is generally used to refer to more formal, organized, and well-established belief systems that are shared by large groups of people. The phrase "spiritual beliefs" refers to more amorphous and usually less established systems of belief. These distinctions are important, but this chapter will focus on their shared aspects—overarching beliefs that explain such basic questions such as the origin, purpose, and proper form of life—and will use the two terms interchangeably.

To simplify language even further, the word "ecclesiastic," which derives from a Greek root that refers to an assembly of citizens, will be used to encompass the terms religious and spiritual in the discussion that follows to emphasize the idea that the central feature of this form of causal logic is that the belief systems that underlie this form of causal reasoning are shared among groups of people and have persisted over a significant period of time. While empathic and empiric belief systems also are advocated by groups of individuals (social scientists and basic scientists, for example), the approaches that they represent change much more frequently and dramatically over time. This discussion is using the term "ecclesiastic belief systems" to refer to sets of central ideas that have been relatively stable over several hundreds or thousands of years. Although the word "ecclesiastic" has two drawbacks—it implies "churchly" to some, and it is the title in English of a book of the Old Testament—it also has the advantages of emphasizing the group nature of the stable belief systems to which I refer and of starting with "e," as do empiric and empathic. It is thus a convenient way to refer to the primary subject matter of this chapter in its broadest and most inclusive meaning—an alternative logic and method of causal reasoning.

COMMON FEATURES

The central feature of ecclesiastic systems is their basis in *given* truth. They begin with a core set of ideas and derive proscriptions from them. These core ideas usually include the knowledge that a basic force exists and that this force has driven and will continue to drive the major events of nature. This contrasts sharply with the empiric and empathic, both of which are focused on the *search for* universals. Said differently, the empiric and the empathic provide methods for identifying universals, whereas the ecclesiastic begins with the knowledge of what they are.

One result of this feature is that many ecclesiastic systems place a major emphasis on "why" questions and explain the purposes behind events that occur in the universe. In contrast, "why" questions are not a major focus of empirical causal reasoning, although they are sometimes addressed in isolated topics such as Darwinian natural selection. The quotation from the noted American astronomer Allan Sandage (1926–2010) at the beginning of this chapter captures the search for causal purpose as an essential goal of ecclesiastic systems.

A second characteristic of ecclesiastic systems is that their core ideas or universals are used to proscribe how people should live. While nonecclesiastic legal and political systems also apply universals to everyday life, their laws are generally recognized as having been produced by humans for the purpose of influencing behavior. Some political systems do rest on general principles based on beliefs about governance or human nature, but they generally do not derive them from final causes unless they are theocratic states and incorporate religious causality into the structure and essence of their function. The emphasis on how individuals should live reflects the idea that ecclesiastic systems begin with truth and then derive how these truths should be actualized or implemented.

Empathic reasoning is sometimes used to identify universals and then to apply them to everyday life, but the direction of reasoning derives from observations of the behavior of humans that are then generalized to universals. In ecclesiastic causal reasoning, the direction is generally the opposite. Some practitioners of empirical methods also seek to derive or describe universal or widely shared behaviors from the study of individuals or groups, but these are formulated as arising from the structure and function of the organism or universe. These are sometimes anthropomorphized into a purpose for doing so (survival of the species being an example), but empiricists do not generally apply labels such as "right," "correct," or "ethically necessary," as happens in the ecclesiastic. This aspect of the ecclesiastic was nicely captured in Sandage's response to a question about the origins of his spiritual beliefs. He stated that he had been

> plagued by the question of purpose . . . and the problem of understanding the basis of ethics and morality. What is the good? Science is value neutral whereas religion is value intensive. . . . I think there has to be an absolute, and the only absolute answer is, what is ethical is what God wills.

The claim that ecclesiastic systems begin with knowledge of ultimate causes does not imply that there are no unanswered questions in the ecclesiastic. In fact, like many empiric and empathic approaches, many ecclesiastic systems emphasize the quest for increasing human understanding as a central focus of their existence. In doing so they frequently posit that human knowledge is incomplete and proscribe methods or paths by which individuals can expand their knowledge. However, most ecclesiastic systems in which this is the case identify the lack of understanding as a characteristic or failing of the human condition; they consider ultimate knowledge and its applications to life to be known and knowable but believe that individual humans must work to discover and understand it. For those empiricists and empathicists who believe that ultimate knowledge is knowable, the quest is to discover it for the first time, not to uncover what is already known to a divine power.

This distinction between the ecclesiastic and other methods of knowing is not new. Plato described similar differences when he contrasted mythology and philosophy, for example. For him, philosophy was knowledge that is accessible to reason and therefore proof, whereas mythology rested on ideas that had been handed down and thus need not be proven. Today, the word "mythology" connotes a system of fanciful gods or stories, but for the ancient Greeks, *mythos* played a role similar to that currently played by those disciplines referred to here as ecclesiastic. Aristotle's identification of a fourth level of causality, labeled "purposive" in chapter 1, also demonstrates that this form of causal reasoning has long been considered distinct.

Another characteristic of the ecclesiastic that contrasts with the other two approaches is the strong emotional component central to many ecclesiastic belief systems. This is not always expressed publicly or openly, and it is not always operative for significant time periods, but emotional involvement is usually emphasized in ecclesiastic systems, and this is often further enhanced by ritual. While it is true that the empiric and empathic realms involve some individuals in a profoundly emotional manner, this rarely persists over long periods of time or operates as group phenomena, particularly for the millennia that they have in ecclesiastic systems. This aspect of the ecclesiastic was also commented upon by Aristotle, who noted that people attended religious events to experience *pathein*, a transcendent state in which the emotions are involved to a degree that differs from the everyday experiences of most individuals.

Δ Δ Δ

In his classic discussion of religion, *The Varieties of Religious Experience*, William James identified five characteristics of the religious life. In his words, they are:

1. That the visible world is part of a more spiritual universe from which it draws its chief significance;

2. That union or harmonious relation with that higher universe is our true end;

3. That prayer or inner communion with the spirit . . . is a process wherein work is really done, and spiritual energy flows in and produces effects;

. . .

4. A new zest . . . takes the form either of lyrical enchantment or of appeal to earnestness and heroism.

5. An assurance of safety and a temper of peace, and, in relation to others, a preponderance of loving affections.

James elevates religion above other endeavors ("our true end") and claims that rituals such as prayer bring about results and depend on an energy or process that has a physical basis. He also makes the claim that religion is a central source of the desirable goals of love and peace. Finally, James highlights the emotional involvement that occurs with religion and believes this distinguishes it from science and empathy. As he puts it, "knowledge about life is one thing; effective occupation of a place in life, with its dynamic currents passing through your being, is another."

As with other distinctions made throughout this book, the differences among the ecclesiastic, empathic, and empiric are not absolute. All three approaches depend on communities of individuals who share similar ideas and often an emotional connectedness to one another. Further, each approach brings together groups of individuals who seek generalities and share a belief in the method being used.

Several other elements are shared by all three of the approaches. Each uses rhetoric to convince others that its formulation of causal logic is convincing. Chapter 9 discussed the central role played by rhetoric in the empathic method, but rhetoric is also used in the ecclesiastic (and the empiric) to convince people of particular ideas and of the relevance of the approach to the important questions of life. Another similarity across the three approaches is that what has been accepted as "truth" has changed over time. However, continuity is much more characteristic of ecclesiastic approaches; adaptation and change have been much more characteristic of the empathic and empiric. This reflects the claim made above that knowledge in the empiric and empathic is incomplete and that these logics provide methods for seeking new knowledge, whereas knowledge in ecclesiastic systems is usually complete but not fully comprehended because of human imperfections in understanding. Thus, even though knowledge evolves or develops or has been discovered over time in all three modes, the reason for these changes is fundamentally different in the ecclesiastic compared to the empiric and empathic.

As mentioned above, the chief characteristic distinguishing the ecclesiastic from the empathic and empiric is that the ecclesiastic starts with given truth. When using the empathic and empiric methods to search for initial causes, one can always ask, "and before that?" Eventually the questioner will get to a point before which no precipitating or predisposing cause can be identified. This has been used to refute the ultimate utility of the empathic and empiric approaches for providing purposive causality, but I prefer to conceptualize this as an inherent limitation of these ways of knowing. The hierarchical question "which is better?" is inappropriate without adding "for what?" This is the essence of the Meister Eckhart quotation at the beginning of this chapter, and it has distinguished his writings from those of prior thinkers who described differences between ecclesiastic/religious and "logical" or "sensory" (experiential) logics.

The claim that there is value to ecclesiastic causal reasoning will be dismissed by those who believe it is a method for glossing over the fact that some things are unknowable. Conversely, those who adhere to the ecclesiastic can claim that its unique ability to describe origins is what distinguishes the religious and spiritual from the mundane. This capacity is viewed with awe by proponents of many ecclesiastic systems and with profound skepticism by disbelievers. It remains a distinguishing characteristic among these three ways of knowing. The distinction is worth keeping because, as Aristotle recognized 2,500 years ago, purposive cause requires approaches and assumptions not inherent in precipitating, predisposing, or programmatic cause.

Each of these approaches—the ecclesiastic, the empathic, and the empirical—is dismissed or rejected by large numbers of individuals. Some individuals embrace two of them but not three. These differences likely have multiple causes, among them the facts that individuals are educated differently; are exposed earlier to the ideas and methods of a given method; have innate skills that predispose them to use one or the other; are born into, appointed to, or hired for positions in which one method is emphasized; or have an intellectual disinclination toward that way of knowing (or at least no inclination toward it). Human history, at least to date, demonstrates that none of these approaches has achieved anywhere near the universal acceptance that would support a claim of supremacy over the other two approaches, but adherents of each believe this lack of universal acceptance or understanding can be overcome by reaching out to those who are not adherents or users of the approach.

On the other hand, all three approaches share the belief that the truths they espouse are available to all. Adherents of each approach often expound the claim that "if only" people were more exposed or more educated in their approach or set of knowledge they would better apply this knowledge and, by implication, be better fulfilled or be better citizens. This claim may seem to apply more often to the ecclesiastic because it overtly deals with principles that guide life, but the literature of the science educator and the humanities educator is filled with similar statements. Whether everyone could or would become versed in the basic tenets of any one or all of these methods and apply them appropriately is itself a question that could be addressed from empirical, ecclesiastic, and empathic perspectives, but I believe that wide variations in exposure, experience, innate capacities of temperament and intelligence, and culture among people make this extraordinarily unlikely. Undoubtedly, individual and group acceptance and understanding of each of these approaches has changed and can change over time, as the Sandage quotations above suggest, but the wide variation in human nature suggests to me that universal acceptance is not achievable, even were it desirable.

As in other chapters in this book, the construct being proposed here is a pluralistic approach that ascribes to each method unique contributions to the construct of causality. This proposal claims that each one is both orthogonal and complementary to the other two. The choice of which approach to use and when to use it rests on the supposition that certain questions are best addressed by a specific method because of its inherent strengths and assumptions and that the others are either less or not applicable. In some instances, the combination of more than one method is best. The primary claim is that the choice of which approach to employ and when to use it is not arbitrary but rather based upon an understanding of each method's assumptions, limitations, and strengths and on the specific question that is being considered.

THE ECCLESIASTIC AND THE EMPIRIC

Attempts to reconcile the ecclesiastic and the empiric date back to the origins of the scientific revolution, and many great and not-so-great thinkers have attempted to use scientific reasoning or findings to support or refute ecclesiastic notions. Galileo and Newton took the tack that their scientific discoveries illustrated the greatness and glory of God, since only a supreme being could be responsible for the regularity and beauty of their discoveries. Whether Galileo truly believed this or merely stated it to escape the wrath of the Church is unclear, since his recantation toward the end of his life of the belief that the sun, not the Earth, was at the center of the solar system seems likely to have been made to placate the Catholic Church and not to reflect his actual beliefs. Newton, on the other hand, ran into no such difficulties and wrote more explicitly about his belief that his scientific discoveries illuminated the power and majesty of God. For Newton, the empiric was a means of demonstrating the truth of the ecclesiastic. Francis Collins, the founding director of the National Institute of Genome Science and director of the U.S. National Institutes of Health at the time this book was published, has made similar claims in his book *The Language of God*. Similarly, the molecular biologist Ursula Goodenough, in her book *The Sacred Depths of Nature*, states that "a cosmology works as a religious cosmology only if it resonates, only if it makes the listener feel religious." She is thus taking those elements of science that she finds particularly beautiful and awe inspiring— emotion inducing—and derives from their beauty and majesty what she calls a "religious naturalism." One learns from the book that she attends a Presbyterian church and thus participates in a classic ecclesiastic setting, but she proposes that the majesty of nature links all humans and that it is this magic that she describes as religious. Similar ideas and sentiments have been expressed by individuals as disparate as Leibniz, Descartes, and Bacon. The nineteenth-century evolutionist Henry Drummond championed the unity of the ecclesiastic model and evolution, and in a 1991 book, *Science and Religion*, John Hedley Brooke notes that "there were Henry Drummonds in every generation who did make connections between their theology and their science."

As noted in chapter 1, one of Galileo's most important contributions to the development of the scientific method and the empiric approach to causality was his suggestion that the empiric approach avoid seeking final causes. Whether this was because he believed that only the ecclesiastic could answer such questions or because he recognized that progress in the empiric required putting aside questions of purposive cause is not clear. What is clear is that much of the success of the scientific enterprise to date can be traced to this insight.

In recent years, brain imaging has been used to examine the biological phenomena that occur in the brain when individuals have what they or the researchers consider a religious experience. In a review of recent writings on the topic of science and religion, the physician-essayist Jerome Groopman criticizes this and other attempts to explain religion in scientific terms: "The cardinal error of neurotheology [is a] mixing of terms and methods . . . in an attempt to confer the former's authority on the latter." Groopman's argument is similar to the one being made in this chapter: the ecclesiastic and scientific approaches rest on different assumptions and use different logics. Ironically, notes Groopman, some of those who claim to find evidence that religious beliefs originate from specific areas of the brain use these "scientific" findings to prove that a supreme being exists. That is, they make the erroneous claim that the occurrence of certain electrical events or blood flow changes in the brain when a person is experiencing what he or she reports as a spiritual experience proves that their beliefs are "real" because brain activity is equated with reality. Others use the same evidence to show that such beliefs are the results of specific types of brain activity and nothing more, that is, brain-generated ideas that have no basis in fact. The error here is that finding an association between brain activity and reported thoughts cannot prove or disprove ecclesiastic beliefs. There is no way to know whether the belief preceded the activity or resulted from it, and furthermore, these methods cannot establish the validity or falsity of the beliefs. Sandage reflected this point of view when he responded to a question about evidence by saying, "I don't think there is any evidence. It is faith, not reason. If there is proof, faith is not needed!" This is the very point made by Meister Eckhart seven hundred years ago.

Evolutionary scientists and the natural history essayist Stephen Jay Gould also advocated for the view that the scientific and the religious are distinct. Gould referred to them as "non-overlapping magisterium" or "NOMA." To him, science is the magisterium that considers what the universe is made of and why it works as it does, whereas religion is the magisterium that considers questions of ultimate meaning and moral value, a distinction that parallels the ideas put forth earlier in this chapter.

Equally impassioned arguments using empirical and empathic logic have been offered as proof that ecclesiastic thinking is untrue and irrelevant. Both Richard Dawkins in *The God Delusion* and Sam Harris in *The End of Faith: Religion, Terror, and the Future of Reason* use a variety of arguments to support the beliefs proclaimed in the titles of their books. These include an inability to prove many religious claims, the scientific implausibility of many religious ideas, and the many harms that have resulted from religious zealotry over the centuries. Dawkins's and Harris's writings are examples of the uses of rhetoric to support ideas that their proponents attribute to empirical reasoning. Their approach thus contradicts Meister Eckhart's dictum that one method cannot be used to denigrate or uphold the centrality of religion in many peoples' lives. Dawkins, Harris, and others sharing their views would certainly disagree, and there is no ultimate authority to decide the question. Thus, one is left with either a reliance on the methods of rhetoric to convince those holding the opposite opinion or the conclusion that both are right but in contexts that require different assumptions and sources for answers.

THE ECCLESIASTIC AND THE EMPATHIC

Many other scholars and thinkers have sought linkages between the ecclesiastic and the empathic. For example, Karen Armstrong quotes Petrarch (1304–1374) as saying "theology is actually poetry, poetry concerning God." Similarly, when William James lyrically suggests that the differences between the empirical and the religious are "in the terror and beauty of phenomena, the 'promise' of the dawn and of the rainbow, the 'voice' of the thunder, the 'gentleness' of the summer rain, the 'sublimity' of the stars, and not the physical laws which these things follow," he is relying on empathic images and linkages to make his point. Thus for James, in contrast to Newton and Goodenough, spirituality is more than wonderment at the diversity or orderliness or complexity of nature; it is the search by human beings for an emotional union with some greater force. James called this "living individualized feelings." For James, it is the deep emotional involvement in the search, what I consider an empathic phenomenon, rather than the intellectual gain that is the primary distinction between spirituality and the empirical.

Somerset Maugham wonderfully captures the development of doubt and the dissolution of religious belief in his semiautobiographical novel *Of Human Bondage*. Maugham depicts the struggle between a loss of religious belief and faith and the struggles to replace it with values derived from other sources. Whether the reader agrees or disagrees with the path taken by the main protagonist, Philip, Maugham's ability to evoke the struggle Philip faces in the story he tells illustrates and highlights the strengths of the empathic approach.

An example of using an empathic argument for the existence of God and religious causality can be found in a statement that came to be known as "Pascal's wager," after the distinguished seventeenth-century mathematician Blaise Pascal, who argued that there were almost no negative consequences of being wrong in accepting God's existence but that the adverse consequences of rejecting God were very significant. He concluded that logic favored accepting God's existence. In my opinion, this is a rhetorically weak empathic argument when compared to those offered by Collins and Goodenough.

ECCLESIASTIC REASONING AS A DISTINCT LOGIC

In the spiritual/religious realm, causal truth is discovered by studying or meditating on existing texts and teachings that contain discovered knowledge. In contrast, the empathic and empiric methods focus on the discovery of unknown truths by using the methods of those disciplines. New knowledge and new applications to current situations can be gained by using any of the methods, but the identification of new knowledge is not a primary goal of the ecclesiastic. The search for a unity with a higher force or forces, the emotional involvement derived from feeling linked to that unity, and the living of one's life in a manner that follows the precepts of that belief system set the ecclesiastic apart from the empathic and the empirical. The emotional satisfaction that is derived from discovery in the empathic and empirical can be powerful and have a significant impact on the lives of many, but it appears to me to be of a different (not better or worse) realm.

A survey published in the distinguished scientific journal *Nature* in 1997 illustrates some of these points. This survey found that 40 percent of U.S. scientists reported a belief in God, a much higher percentage than that found in European scientists. The authors of the commentary expressed surprise at this percentage, likely because it conflicted with their own beliefs, but these numbers can be used to bolster whatever point of view a person is trying to support. On the one hand, many individuals who consider themselves scientists view religion as a powerful and alternative but not equivalent method of knowing. On the other hand, a majority of scientists view traditional Abrahamic religion as peripheral to their lives. The question of whether religion and science are distinct was not asked directly, but the results of the survey suggest that almost all scientists discriminate between them.

CAUSALITY IN THE HINDU AND THE ABRAHAMIC RELIGIONS

The following paragraphs very briefly consider the conceptualization of causality in two formal religious traditions, Hinduism and the Abrahamic religions. Generalizing about these or any broad tradition unavoidably overlooks many details and specific concepts that meaningfully affect how causality is viewed within that tradition, but a brief discussion does provide an illustration of what has been discussed in this chapter. It particularly emphasizes the early origins of these ecclesiastic systems in quite different civilizations, their persistence over time, and the presence of ideas that seem startlingly modern.

The Hindu Tradition

Hinduism approaches causality from a cyclical or circular rather than linear point of view. It conceptualizes history as inherently repetitive and posits neither a beginning nor an end of time. This pattern derives from the *dharma*, a given law of the universe. The result of dharma is that cause and effect are intertwined. For example, dharma both drives human action and results from it. Nevertheless, humans have initiative and therefore responsibility for their actions, and it follows that they have causal power. An example of the lack of a beginning is that the Vedas, the sacred texts of Hinduism, were written neither by a god figure nor by man. Rather, they just "exist." To a Western mind this has little or no meaning, but to the Hindu it speaks of the timelessness of the Vedas' words.

Many Hindus are vegetarian, and the Vedas offer support for this. Reasons include nonviolence toward living creatures and the belief that, because of karma, harm inflicted to others will be later experienced by the person inflicting the harm. This is an example of a moral and behavioral guide that derives from this ecclesiastic system.

Like many spiritual and religious systems, especially those that have existed for a long time, Hinduism has many varied paths or sects, and these offer different approaches to what seem like very basic issues. For example, the Sankhyu school embraces a dualistic approach that distinguishes between physical and spiritual causality and between mind and body causality, whereas most others who consider themselves Hindu do not make such distinctions.

This variation among the schools of Hinduism and across the multiple sects of many ecclesiastic belief systems might seem to contradict the generalization made earlier that spiritual/religious systems change relatively little over their history. However, many of the central tenets of Hinduism have been intact for several thousand years, and the differences are both quantitatively and qualitatively fewer than the changes undergone by what is now called science over the same period of time.

On the other hand, some aspects of Hinduism seem startlingly modern. Examples include ideas such as the "reversibility" of time, the interchangeability of cause and effect, and the relativistic nature of time (depending on the observer rather than an existing absolute standard), all reminiscent of ideas that are central to relativity theory and quantum mechanics. Many of the early quantum mechanics theorists were aware of these similarities, and the popular press has embraced these ideas in books like *The Dancing Wu Li Masters* and more recently *The Elegant Universe* by Brian Greene. Some writers have taken the similarities between these ancient traditions and modern physics to be an indication of the unity of the ecclesiastic and the empirical, a view discussed and rejected earlier in this chapter. More instructive, in my opinion, is the fact that some of the very ideas that make quantum mechanics and relativity theory seem counterintuitive to many Westerners are embraced by many millions of Hindus. This demonstrates the limitations of relying on "intuitions" or "obviousness" as proof of belief about the world or as reasons for rejecting other constructs, and it emphasizes the value of studying ideas from other cultures when seeking to determine what is "natural," "obvious," or "necessarily" true.

The Abrahamic Tradition

The three major religions of the West and Middle East, Judaism, Christianity, and Islam, share a belief in a single God from whom the tradition derives. This God is all knowing and everywhere—omniscient, omnipresent, and omnipotent—and in that sense is the ultimate source of causality. Since ultimate cause derives from one source and from one point in time, causality is a linear process in the Abrahamic traditions.

In addition to sharing a belief in a single God from whom all emanates, these three religions share a common human founder in the prophet Abraham. Islam traces its origins to his son Ishmael; Judaism and Christianity trace their lineage through his other son, Isaac.

Great variation exists not only among these three religions but within the many sects that have formed within each. For example, some sects perceive God to be the ultimate source of all actions; others conceive of God as the initiator of a process that has operated on its own and thus not under His control since the initiation of time; the latter consider God a more passive observer of the actions of humankind and the world as opposed to the cause of all worldly events and human behaviors. These are quite different notions of causality, even though they share a basic belief in a single source. Each of these traditions shares the belief that a time will come when evil will disappear and a new set of principles will drive the interactions among the peoples of the world. A very similar idea exists among the Hindu traditions. Predictions such as these can be found in the science of cosmology (the universe is now thought to be expanding forever rather that expanding and contracting periodically) and in many utopian narratives, but in science the current structure of the universe is conceptualized as resulting from natural forces set in motion at the beginning of time and not under the control of humans, while in the narrative, political traditions they result from the actions of humans rather than from a being who has the power to bring about such a state. However, the existence of such similar ideas convinces those who seek to minimize differences among these approaches to acknowledge that the differences are inconsequential or exist primarily in the minds of those who would overlook the obvious. The source of either belief, however, is in the realm of conviction or narrative rhetoric and not accessible to other forms of proof.

The centrality of law is one tradition that these three religions also share. All recognize Moses or Musa as having received a set of rules from God and derive detailed proscriptions for behavior from them. Here again there is significant variation, with some teachings emphasizing peace and others aggression, some sects forbidding dancing, others seeing it as a behavior that can lead to a closer relationship with God. What is shared, though, is the central idea that God is the source of a set of basic laws that all humans should follow.

In his wonderful book *The Science of Conjecture*, James Franklin traces the development of the modern concept of probabilistic reasoning from attempts of Judaic scholars around 100–200 CE to solve causal questions when the details of a situation made the application of these laws difficult to decide. (As is noted in chapter 3, the Roman Digest, another early source of English common law, took a similar approach and emerged about the same time). These opinions were collected in works called the commentaries or Mishna and the Talmud, and they are one source of the common-law tradition of relying on precedent. The Talmud's presentation of an approach to causality that identifies graded degrees of causal responsibility can be traced back to the need to solve ethical and legal disagreements long before probability was developed as a mathematical approach in the seventeenth century.

CONCLUSION

The search for ultimate cause has ancient roots and seems to have been part of the human psyche at least as far back as permanent human records exist. Many cultures and groups of people have come to believe that a single entity, group, or essence, often intangible but knowable through study and experience, brought about the physical universe in which we live and provided guidance on how lives should be lived. This idea, expressed in the Rig Veda, Sumerian tablets, Babylonian myths, and Genesis, remains a basic and essential belief of much of the world's population today. While there are many variations on this theme and other models of spiritual thought that follow different patterns, that this idea can be dated to the earliest human writings and retains great causal power today for many people demonstrates both the centrality of the idea of causality to human thought and the likelihood that this form of causal reasoning will continue to be both widely held and irrefutable for the foreseeable future. Some will continue to see it as the primary source of causal knowledge, others as one of several sources, and still others as an outmoded and even destructive approach. The view presented here is that spirituality/religion can most powerfully and usefully be understood in the modern era as a model that complements the empathic and empirical approaches. This thesis can be debated but, without an ultimate arbiter of truth, we have only the tools of rhetoric to convince others of the ultimate truth of our ideas.

The theologian Meister Eckhart came to a similar conclusion centuries before what we now recognize as the scientific method emerged. His claim, quoted at the beginning of this chapter, that neither science nor logic can address the questions raised by those who seek to know purpose seems unassailable. This question can be rejected as irrelevant, unanswerable, or meaningless, but the fact that purposive beliefs have arisen independently many times throughout human history, their persistence in many if not all cultures throughout history, and the strong hold that such beliefs continue to have today throughout the world argue for its being a distinct approach to causality. Aristotle made a similar argument 2,500 years ago when he identified purposive causality as a distinct approach to the construct.

11

SEEKING THE WHY OF THINGS
The Model Applied

The promise of inevitability is so enticing because it claims we
don't need to know the detailed rules.

—Duncan Watts

In retrospect we can make sense of life, but life must be lived
forwards.

—Søren Kierkegaard

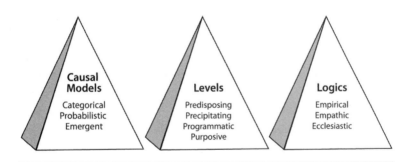

Chapters 1 through 10 have established a conceptual model for
approaching questions of causality, and they used examples to illustrate and clarify these ideas. In this chapter, the emphasis will be
reversed. Six specific issues—the emergence of HIV/AIDS as an
epidemic illness worldwide, causality in U.S. law, evolution as a causal
concept, Alzheimer disease, human aggression, and the etiology of
depression—will be used to discuss the three-facet model and illustrate both its utility and limitations.

HIV/AIDS

In 1981, doctors in Los Angeles reported a mysterious illness that had afflicted five young males. All were homosexual and had been healthy until they developed a progressive, debilitating illness associated with unusual infectious diseases, a set of circumstances suggesting that their immune systems were dysfunctional or suppressed. The illness was labeled "gay-related immune deficiency," or GRIDS. Other urban-based physicians quickly reported caring for individuals with striking similarities: most had been previously healthy, most were males who reported having had same-sex partners, all suffered from debilitating infections, and many had died. Most of the individuals with the disorder who did not report having same-sex partners had either used illicit substances that were injected or had received blood products as part of their medical care. The disease was relabeled "Acquired Immune Deficiency Syndrome" (AIDS) in English-speaking countries.

The clinical features of the patients and the fact that most had same-sex partners, injected illicit drugs and therefore were likely exposed to needles used by others, or had received blood transfusions pointed to an illness caused by an infectious agent that spread via contact with body fluids such as semen and blood. The uniqueness of the symptoms and laboratory studies suggested that this was a "new" disease, that is, a disease unknown to modern medical practitioners. The only available treatments were those that targeted the infectious diseases or cancers that occurred as a result of an ineffective immune system. Most individuals had repeated bouts of both common and rare infectious diseases, many developed dementia, and many died in a debilitated, cachectic state. The illness was myriad and devastating, and by the time it became widely recognized, it had spread across continents and moved into more rural areas. Within several years it was a worldwide epidemic illness.

In 1983, a scientific group led by Luc Montagnier at the Institute Pasteur in Paris reported that individuals with AIDS were infected with a previously unknown virus that was not present in individuals without AIDS. They called it LAV (lymphadenopathy associated virus). The next year, Robert Gallo and coworkers at the National

Cancer Institute in Washington reported the isolation of a virus they called HTLV-III. These viruses were soon shown to be identical and present in all individuals with AIDS (step 1 in Koch's postulates; see chapter 1). Each of the research groups claimed discovery of the cause of AIDS, and the virus was renamed HIV (human immuno-deficiency virus) as a partial resolution to this naming controversy.

Scientists studying the virus identified it as a retrovirus, a virus whose RNA directed the formation of DNA. Drugs previously developed to treat such viruses were shown to be partly effective, but resistance to them developed very quickly. Over the next decade, scientists identified many features of the virus's biology and through that knowledge produced more effective pharmacologic therapies. Eventually, a complex regimen of several drugs, called HAART (highly active antiretroviral therapy) was shown to impair dramatically the virus's ability to reproduce and thereby greatly mitigate its adverse health effects by preventing its spread throughout the body. However, these drugs are unable to eliminate the virus from the body because the virus is able to "hide out" in cells. HAART has dramatically changed the course of HIV/AIDS, as the disease is now called, because the unimpaired immune system is able to prevent most of the debilitating infections that previously characterized the disease. It has thus enabled many individuals to remain healthy as long as they can afford and are willing to take the drugs.

Even before the agent causing the disease was discovered, though, it was clear that many cases of the disease could be prevented by interrupting the sharing of body fluids, by using condoms and clean needles and by screening blood donors for risk factors. Some critics have claimed that this strategy was not disseminated quickly enough, but even today, when knowledge of how the disease is spread is widely known, such preventive steps are still ignored.

The discovery of the virus also led to the development of a test to detect its presence in blood. As a result, all donated blood could be screened, making the replacement blood supply safe. (Again, a lack of rapid dissemination in several countries led to infections in people suffering from hemophilia, who require transfusions of a factor in blood that promotes blood clotting, and others receiving blood transfusions.) Today, HIV/AIDS exists throughout the world. The number of new cases has been declining in many places, but its spread has not been eliminated. In some developing nations, treatment is still not widely available or affordable, and in countries where it is, new cases continue to occur because preventive strategies have been implemented poorly or have not targeted the specific predisposing causes that are prevalent in that area.

The development of antiretroviral drugs has dramatically changed the disease from an inevitably fatal illness to one that can be controlled but not yet cured. Treatment of individuals exposed to the virus within forty-eight hours can prevent infection, and the treatment of pregnant women infected with the virus at the time of delivery can prevent transmission of the virus to their child. These tremendous advances came about through empirical research that depended on the knowledge that the virus is the precipitating cause of the disease.

However, neither the knowledge of how the disease spreads nor the effective therapies developed in the last decade was successful in stopping the spread of the disease to all corners of the earth or led to lower rates of new cases in many parts of the world. Clearly, transmission must be interrupted if the epidemic is to be halted, but it will take an understanding of the behaviors of individuals and groups of individuals to do so. For example, male circumcision is associated with a lower risk of virus spread, and having multiple sexual partners is associated with higher risk. Changing the patterns of these risk factors is a significant challenge.

The Three-Facet Model Applied

The *precipitating* cause of HIV/AIDS, the HIV virus, was discovered just three years after the disease was first identified. Without the virus, the illness does not occur, and when the virus is present in a human, the immune system is almost always suppressed or becomes suppressed over time. These facts fulfill the first of Koch's criteria for causally linking a pathogen to a disease. Since no other factor (drug, behavior, gene, infectious agent) has been found to be universally associated with the disease, there is universal agreement among experts that the virus is the cause of AIDS. Those few individuals who have claimed otherwise have not put forward any evidence to support their claim. Treatments have been developed that can significantly decrease the amount of virus in the bloodstream of an individual and presumably throughout the body. This partially fulfils the second of Koch's criteria, which states that removing the offending agent resolves the symptoms. Many scientists are attempting to develop a vaccine that would prevent susceptibility to the virus and prevent the disease from occurring. Some experts believe that vaccination could be as successful as the smallpox vaccination program and lead to the virus's elimination from humans. However, the difficulty of eradicating other viral diseases such as measles and the failure to eliminate other sexually transmitted diseases such as syphilis and gonorrhea, for which effective treatments have been available for decades, indicate the great difficulties that would have to be overcome for this to be achieved, even if the biological challenges of developing a vaccine are surmounted. Understanding unique aspects of the complex causal web of each disease will thus be necessary to make further advances in prevention and treatment.

Predisposing causes of HIV/AIDS were identified even before the virus was isolated: behaviors and actions in which body fluids were shared between individuals—either vaginal, anal, or oral sex; the use of a needle by more than one person; the injection of a contaminated fluid or transplantation of an infected body part by medical practitioners; and the vaginal birth of a child to an infected mother. These actions increase the likelihood that the disease will spread from one individual to another, but they cannot lead to the disease unless the virus is present. Since interruption of them could prevent the virus and thus the disease from spreading, they are essential parts of the causal chain.

One extraordinary aspect of HIV/AIDS is how quickly it spread around the world. Factors contributing to this included the ready availability of air travel because of its low cost and universal access, the prevalent use of injected recreational drugs, social settings and changing attitudes that have increased the average number of sexual partners one has, extensive use of blood products by medical professionals, and the failure of medical practitioners to follow precautions known to prevent the spread of infection, such as the wearing of gloves and the single use of needles and syringes for injection. While each of these factors contributed to the epidemic spread of the HIV/AIDS around the globe, the disease would likely have become established without them, just more slowly. Their role as enablers is what makes them predisposing causes.

Stigma about the illness was and remains another contributor to the unrestrained spread of the HIV virus since it prevented some individuals with symptoms from seeking care or mentioning that they were ill to sexual partners, physicians, or others. Some claim that a slow response by public health and elected officials also contributed to its quick spread. These cultural factors were and remain *programmatic* causes. They are a part of the fabric of the environment in which the virus became established and fostered its spread, by increasing the likelihood that the precipitating and predisposing causes could operate. Social networks consisting of individuals who share certain behavior patterns also contributed to the spread of the virus. In the West, the disease first struck men with same-sex partners because in some cities they formed highly interactive groups. Likewise, those sharing needles are likely to interact with one another and further spread the virus. As researchers such as Barabási showed, a few highly connected individuals ("hubs") can have major effects on a whole network. A few individuals with many sexual partners, whether they are opposite- or same-sex individuals; a single infected individual whose blood components were mixed with those of others to provide a treatment given to many individuals with hemophilia; or a single saline bottle that was contaminated by a needle that had been used multiple times by medical professionals or substance users can spread the virus to many others, each of whom might then pass it on to others. Understanding that highly connected "sites" play a role in the spread of the condition provides an understanding of cause that differs from the precipitating and predisposing factors discussed previously.

The logic of *empirical* scientific study as applied by epidemiologists, virologists, pharmacologists, and infectious disease specialists led to the conclusion that HIV/AIDS is an infectious disease, that it is spread by the sharing of body fluids, and that the causal agent was a specific and previously unknown virus. The evidence for this was that (a) the disease clustered in certain groups of individuals (males who had had sex with males, substance abusers who had used drugs by injection and likely used needles previously used by another substance user, recipients of blood transfusions or other blood products), (b) that what these disparate groups had in common was an interchange of body fluids (semen, blood), and (c) that many of the associated infections and cancers suffered by those with the condition were known to be caused when the immune system was suppressed. As a result of this reasoning, effective suppressive treatments of the HIV virus were developed in an extraordinarily short time (although not fast enough for many victims) because some antiviral drugs had been previously developed. An understanding of the mode of transmission also led to programs that promoted protected sex, provided alternatives to needle sharing, screened blood products, and led to the use of gloves by health professionals to block transmission between them and those they touch. A decade later, a more detailed understanding of the life cycle of the virus led to the development of more effective treatments. Thus empirical causal logic played the central role in identifying the cause and controlling treatments for the disease. These empirical data also shaped the development of prevention strategies. Individuals donating or selling blood were screened for the risk factors, and those with risk factors were excluded from providing blood. Tests were developed to identify the virus in blood products and infected blood. Strategies to prevent the spread of the virus by stopping the exchange of semen were also instituted. These include encouraging the use of condoms and discouraging casual sexual encounters.

Empathic logic makes plausible the contributions to disease spread of easy access to air travel, changing attitudes that increased the sharing of needles used to inject drugs that influence mood and behavior, and changes in sexual mores. Even today, in spite of universal knowledge of how the disease is spread and how it can be prevented, new infections continue to occur. Some of the contributors to this distressing fact are not amenable to empirical testing but are highly plausible. These include age-specific developmental factors (such as feelings of invulnerability and rebellion), personality traits (some individuals are by nature more likely to take risks when adverse outcomes are possible and known), the rejection by some governments and individuals of public health measures such as the distribution and use of condoms, the lack of medications for HIV-positive pregnant women at the time of delivery because of cost and inaction, the power of the sexual drive, and the addictive nature of some injected drugs that pushes some individuals to takes risks. These factors are highly plausible contributors to the establishment and spread of the epidemic, and their role seems indisputable to me. But while there is some factual evidence backing up these claims, for example, an association between injectable recreational drug use and acquiring the virus, the hypothesis that this can be attributed to changing attitudes and socioeconomic variables depends primarily upon the narrative power of the claim.

Where did the disease come from? The similarity between the HIV virus and a virus found in African nonhuman primates, the simian immunodeficiency virus (SIV), and the evidence that the first cases emerged in Africa in the 1950s suggest that the virus passed from monkeys to humans. A plausible theory is that hunters cleaning monkeys killed for meat were infected through cuts or open wounds on their hands and that the virus mutated to a form that could infect humans. This narrative seems very plausible and puts together what is known about the SIV and HIV viruses, the fact that the earliest cases likely occurred in sub-Saharan Africa, that the virus is spread by blood, and some evidence that there were cases in the 1950s. The "sudden" emergence of the disease and its explosive spread in the early 1980s can be understood as the concatenation of these various factors. No one predisposing or precipitating cause explains why the epidemic occurred or why it happened when it did, but when considered together, a plausible, coherent, comprehensive, and powerful narrative emerges. Furthermore, other factors have been suggested that are not universally or widely accepted. For example, Craig Timberg and David Halperin suggest in *Tinderbox: How the West Sparked the AIDS Epidemic and How the World Can Finally Overcome It* that the colonial system and its remnants contributed to the emergence of the disease.

HIV/AIDS also illustrates the coexistence of multiple causal models, each of which adds explanatory power. The causative nature of the HIV virus is *categorical*. Without the virus, the disease cannot be present; when the virus is present in an individual, the likelihood that the person will develop the disease HIV/AIDS unless an antiviral treatment is instituted is very high. Furthermore, treating the disease with specific treatments interrupts its relentless progression and has dramatically reduced rates of severe disability and death. These two facts fulfill two of Koch's postulates. The requirement that the virus be spread by body fluid sharing is also categorical.

Many of the predisposing factors of HIV/AIDS are *dimensional*. The more sexual partners an individual has, the greater the likelihood of exposure to the virus and the greater the likelihood of developing HIV/AIDS. Many of the risk behaviors are dimensionally distributed in the population. For example, the personality trait of risk taking is dimensionally distributed in the population. Scoring higher on this trait is associated with a greater likelihood of developing substance use disorder, having unprotected sex, and rejecting other known procedures that reduce risk, all behaviors that increase the likelihood of exposure to the virus. Epidemiologic studies have also shown that the spread of HIV/AIDS can be modeled by understanding the social interaction patterns of high-risk groups of individuals such as men with same-sex sexual partners, men who frequent prostitutes, and intravenous drug users. These interaction patterns help explain the explosive *nonlinear* growth of the epidemic and identify targets ("hubs") for prevention and early case detection.

In spite of the above discussion, it is reasonable to ask whether the three-facet model improves understanding of the disease and, more importantly, improves efforts to prevent and possibly eliminate it. Clearly, until therapies that either eliminate the virus by killing it or prevent it from becoming established by inducing an immune response (a vaccine)—that is, target the categorical cause—are available, the disease will continue to spread *unless* there is universal implementation of strategies that target the predisposing causes— shared use of needles, unprotected sexual intercourse among individuals with multiple sexual partners, failure to provide antiviral treatment to HIV-positive women about to deliver a baby, and the practices of medical professionals that expose multiple individuals to blood and blood products. The application of the three-facet model makes it clear that the elimination of this scourge will require both the development of better biological therapies that target the causal virus *and* the implementation of effective strategies to eliminate or at least greatly lessen the individual and network factors that predispose to exposure of the virus. The failure to eliminate syphilis and gonorrhea in spite of the availability of effective treatments for more than sixty years suggests that the development of a drug that eliminates the virus from the body will not be likely to stop the spread of the disease on its own. The development of public health strategies can and have lowered rates of new cases, but the virus continues to spread. The failure to eliminate other sexually transmitted diseases and to stem the tide of substance abuse disorders despite decades of trying suggests that HIV/AIDS will be difficult if not impossible to eliminate. The development of a vaccine that can prevent the disease would undoubtedly be a great step forward, but understanding that many of the predispositions to developing the disease are in the realms of human behavior makes it clear that no single strategy will be effective. Advocates of each of these approaches often paint their approach as the way forward. The three-facet model suggests that progress toward eliminating or at least diminishing the suffering caused by HIV/AIDS will require better understandings of the categorical, predisposing, and nonlinear causes and the development of strategies that target factors in each of these categories. Thus, the three-facet model emphasizes the need for continued efforts on multiple fronts and provides a framework on which such a strategy can be explained and guided. The complexity of the three-facet model is a challenge, but it allows a way to counter those who dismiss public health strategies, discredit the importance of disease biology, or believe that condoms, programs to encourage monogamy or single sexual partners, or medications alone are the answer to stopping the epidemic. Any one of these is doomed to failure because it is only one aspect of a complex causal chain.

The three-facet model also helps explain why it has been hard to make headway against some of the predisposing and network causes and why education alone has not stopped such predisposing behaviors from happening. The design of public health interventions must recognize that no single approach can eliminate or mitigate the spread of the virus and that any effort to attack the predisposing factors will face many challenges. Furthermore, even though a large effort is being put into the development of a vaccine and drugs that can kill the virus, the difficulty encountered thus far in addressing the predisposing causes suggests that progress toward disease elimination will be stymied unless the dimensional and narrative predisposing issues are addressed.

The limitations of the three-faceted approach to causality are also illustrated by the example of HIV/AIDS. Some public officials and individuals have beliefs that are not swayed by data that are empirical or narratives that are empathic. Those who see the disease as a punishment, as a plague introduced by some groups to inflict debility and death on others, or as an illness for which individual responsibility for behavior has no part because it is caused by a virus will be unconvinced and probably not amenable even to discussing the issues from such a complex point of view.

THE LAW

The concept of law encompasses two broad ideas: (1) groups of people need rules to define conduct acceptable to that group, and (2) the rules and procedures used by governmental units to determine responsibility for events or actions, resolve disputes, and determine the type and extent of punishment to be meted out once responsibility has been established should be explicit. Formal laws have existed for at least four millennia, when the Sumerians, a group inhabiting what is now southern Iraq, carved them in stone. Thousands of widely differing systems of law have been established by governing units since then, but the determination of causality is a central feature of all of them. The system of law in the United States builds upon the English system that was in force in the colonies before America became an independent country (except in Louisiana, whose laws derive from the Napoleonic Code).

The essential assumption of the American legal system is that individuals and group entities (corporations, governments, organizations) act as agents with free will and therefore bear responsibility for their actions. It divides wrongs into two broad categories: criminal and tort.

In the criminal law, the definition of a wrongful act is identified in legislatively approved laws. Illegal acts range from walking across the street at an improper place or time (jaywalking) to killing with intent. Several general principles underlie this approach to criminal conduct. First, it is society, through the legislative branch, that identifies the acts it considers illegal. Second, since the written law cannot address the nuances of every circumstance, prior cases (precedents) are relied upon if an issue arises during a legal proceeding that is not directly addressed in the law or about which there is disagreement.

When a verdict is reached in a criminal legal proceeding, it must yield a binary (categorical) decision of guilty or not guilty. Yet absolute certainty about causal responsibility is not required. Rather, causality must be shown to have been established "beyond a reasonable doubt," a probabilistic standard that is the equivalent of "very highly likely." While legal scholars and courts strongly resist putting an actual number to this probabilistic language, it is generally acknowledged to be in the range of the most commonly used standard in science—95 percent likelihood. In science, this means that the likelihood that the results are not attributable to chance are nineteen out of twenty (95 percent) and that the likelihood of the finding occurring by chance is one in twenty, or 5 percent.

The tensions discussed in chapter 2 between the categorical (binary) and the dimensional (continuous) are illustrated in this dual approach: society (as expressed through its governmental organizations) has a need to punish perpetrators of acts that it has defined as illegal. This requires a legal system that determines innocence or guilt in a "yes/no" determination. Yet custom, formalized as precedents dating back many hundreds of years, has recognized that a requirement of 100 percent certainty is impossible and therefore impractical; in the real world of decision making, one can almost always find some fact or piece of information that undermines the ability to be absolute. Even when a person admits guilt there is the possibility that the confession was coerced. In the United States, concern about this possibility is even raised to the level of a constitutional protection against being required to testify against oneself. Doubt can exist even in the face of very strong evidence. For example, even if physical evidence, for example, DNA that could only have come from the alleged perpetrator, is present, the possibility that it was placed there by someone else can always be raised. In spite of these Humean objections that absolute proof or surety is not possible, society has determined that a method for ascertaining responsibility for illegal actions is necessary, and the legal system is the result. Beginning with the framers of the U.S. Constitution, American legislators, jurists, and legal scholars have attempted to construct, through stated legal principles and specific laws, mechanisms to increase the likelihood of accurate decisions and lessen the likelihood of incorrect ones.

The central features of criminal law relevant to a discussion of causality rest on several assumptions: truth exists and can be determined; absolute certainty is too high a standard for determining cause; approaches can be built into the legal system that increase the likelihood of a correct verdict; the success of the system requires that the populace and the experts agree on the basic tenets of the system; wrong decisions can be made and therefore a system of appeals is needed to reconsider decisions; and a yes/no decision must ultimately be made.

In the United States, a second element of the legal system, tort law, addresses wrongs that occur from negligence. This is defined as a failure to follow community-established standards of conduct. A tort claim is instituted by an individual rather than by the state. It addresses questions such as whether a person has harmed another by doing something (digging a hole without surrounding it with a barrier, for example) or whether a corporation made a defective product and whether this product caused a harm to occur.

The central feature of tort law is that parties responsible for adverse outcomes are responsible for redressing the harm done, often by making a monetary payment to the aggrieved. From the point of view of causality, a major difference between tort and criminal law in the United States is that the level of certainty required to decide a tort case is "a preponderance of evidence," meaning "more likely than not." Again, legal scholars reject applying a numerical value to this phrase, but it essentially means that the likelihood of truth is greater than 50 percent. This is much less stringent than the "beyond a reasonable doubt" standard for criminal proceedings. The clear difference between the two standards relates to perceived differences in the nature of the transgression. From the vantage point of interest in causal reasoning, the key point is that the United States has developed two different legal standards for establishing causation based on a perceived difference in the strength of evidence needed to establish causality.

This difference between criminal and tort law extends to important procedural matters. In criminal proceedings, defendants cannot be required to testify against themselves, a protection against self-incrimination that is granted in the U.S. Constitution's Bill of Rights. In tort cases, in contrast, defendants can be required to turn over all relevant materials and to testify, even if a truthful answer might harm them.

Not all tort cases are decided with a yes/no decision. Both sides can agree to a plea in which a fine is paid or some other step taken to address the problem but in which no admission of responsibility is made. Here the causality question is sidestepped, but redress is nevertheless provided. This illustrates the fact that the law has the dual purposes of establishing causality and establishing an appropriate punishment or remedy.

Several other aspects of U.S. law express a more graded approach to causal responsibility and to decision making. Acts that were premeditated are handled more harshly in the criminal system than those judged to have occurred in the moment. An example is the difference between first-degree murder and manslaughter. In tort cases, acts that are deemed to be particularly egregious or repetitive can result in more onerous punishment, for example triple damages, than those that are not. Thus, punishment can be graded according to the circumstances that preceded it, contributed to it, or surrounded it. Judges and juries can decide that extenuating or mitigating circumstances can lessen the responsibility for cause. Current debates about whether a prior history of physical or emotional abuse should temper the responsibility of a person who harms a spouse, parent, or former supervisor and about whether people with mental retardation (now referred to as intellectual disability) should receive the same sentence as a cognitively normal person reflect the tension that results from a model in which causality is recognized as a dimensional construct.

Another example of a graded approach to decision making is the use of panels of judges to make decisions. These panels are usually made up of an odd number of individuals who "vote" on the outcomes. The decision requires a majority (a tie lets the prior decisions stand). Many absolute opinions of the Supreme Court in recent years have resulted from 5–4 votes.

To summarize, causality in the legal sense is defined by the culture. In the United States, determinations of causal responsibility rely on principles and procedures established in the U.S. Constitution and its amendments, through the legislative process of passing laws in various jurisdictions, and by the accumulation of precedent cases that clarify issues not explicitly addressed legislatively or in the Constitution. Over time, the acts considered right and wrong can change, as can the definitions of causal culpability. This can come about through overt legislative changes to the Constitution and law or through changing interpretations of the Constitution and legislated acts. Even though such changes usually occur slowly, causality in the legal sense is a fluid construct that has and will continue to change over time.

The Three-Facet Model Applied

For cases that come to trial, the primary decision to be made by a judge or jury is whether the accused was the *precipitating* cause of the event in question. They must come to a categorical yes or no. However, if the defendant is guilty, consideration is also given to *predisposing* circumstances in defining the crime and its punishment. For example, the differences among first-degree murder, second-degree murder, and manslaughter include whether there was premeditation and whether there was intent to kill. These are dimensional issues, as they define a range in the severity of the crime. Other examples of dimensional elements are whether mitigating prior elements were present that limited responsibility (an abusive upbringing, for example) and whether the accused had limited cognitive capacity and therefore could not fully accommodate his conduct to the law. In tort law, provision is made as to the grievousness of the crime and whether there were repeated or isolated breaches in the standard.

Programmatic cause operating in the legal context include mitigating environmental or societal circumstances and the ascription of responsibility for an adverse outcome to groups, collections of individuals, or legally defined entities such as corporations rather than individuals. *Purposive* cause relates to the intent of the perpetrator (for example, unintended death is called "manslaughter"; intended death is called "murder"), and this can directly affect the decision of responsibility (if a person is charged with murder and a jury decides that the death was unintended, it could find the person "not guilty" even if it concluded that the individual caused the death).

Categorical logic is required in every case in which a final legal determination is made since a decision of causal or not causal is a necessary outcome, even when mitigating circumstances are present. Categorical decisions include guilt/nonguilt, responsible/not responsible, and liable/not liable. It is noteworthy that the law is based upon the premise that causes exist and that individuals and organizations can be identified as causal agents. As noted in chapter 1, this is a necessary assumption in examining the concept of causality. The law provides an example of a whole system built upon this assumption, and many examples of how causal logics are applied in a variety of situations. *Dimensional* and *nonlinear* logic are applied in many circumstances as well. This ranges from the issue of the severity of the act (first- or second-degree murder, voluntary manslaughter, involuntary manslaughter) to the determination of mitigating circumstances that reduce but do not abolish the penalty, to the degree of confidence with which the evidence is required to support a decision ("beyond a reasonable doubt" being the highest level, a "reasonable degree of certainty" being a middle level, to a "preponderance of evidence" being a lower standard). The *nonlinear* aspect also includes the recognition that some evidence might support a causal role, but if the threshold is not reached then a finding of not guilty is the appropriate outcome. Interestingly, Leibniz, one of the developers of the calculus, cited law as a major source of his ideas about probability and likelihood.

Evidence is ideally *empirical*, that is, of a type and quality such that any jury or judge would come to the same finding. However, the linkages among pieces of evidence and the inclusion of concepts such as motive is *empathic*, just as it is in the sciences and narrative fields. In the United States, ecclesiastic reasoning should not enter the deliberation. The issue of evidence quality will always be a challenge. One reason is the Humean dictum that absolute surety is not possible. Additionally, though, humans are fallible, and the evidence they evince is dependent on characteristics that cannot be guaranteed such as honesty, openness to alternative interpretations, and a lack of conscious or nonconscious biases.

Two events in recent years illustrate the challenges facing the validity of empirical evidence. First, there have been hundreds of reversals of verdicts since DNA "fingerprinting," a technique that can reliably determine whether the DNA on a piece of evidence is from the suspect or someone else, became available. Second, it has been shown that using photographs to identify alleged perpetrators often incorrectly identifies a suspect. The potential fallibility of evidence places pressure on the system to redefine and scrutinize continually the adequacy of empirical evidence that links a suspect to a crime. Built-in safeguards such as the citation of multiple lines of evidence, reminiscent of the approach suggested by Bradford Hill, will always be necessary for the system to meet its stated goal of accurate findings.

HEREDITY AND EVOLUTION AS CAUSAL CONSTRUCTS

Can heredity be said to have a cause? This is a big question, in the league of "Why is the universe structured the way it is?" Genetics, the study of heredity, had its scientific origins in the work of a Bavarian monk, Gregor Mendel, in the 1860s, although humans had been modifying the genetic properties of crops and animals for millennia by crossbreeding livestock and plants with desirable characteristics and producing offspring that were more likely to have the desirable traits. Mendel's experiments demonstrated two facts now taught to all middle-school science students: first, heredity results from the passing on from one generation to the next of units of genetic information (later named genes), and second, that each parent contributes a single copy of each gene to their offspring and that it is the pair of genes contributed by each parent that determines the heredity makeup of the offspring. Mendel's experiments with peas demonstrated that the units of heredity operate in one of two modes. In the dominant mode, a single gene from either parent determines the characteristic or the trait in the offspring, irrespective of the gene that is inherited from the other parent. In the recessive mode, two identical copies of the gene are necessary for the trait to be expressed. The power of this finding was not appreciated until the beginning the twentieth century, and for almost the entire century, this Mendelian model of inheritance drove the field of genetics.

However, by the 1990s it became clear that many diseases and characteristics of organisms do not follow the Mendelian mode of inheritance. In some instances, multiple genes interact to cause a trait or disease, but in other circumstances the mechanism by which genes influence inheritance is either much more complicated or not yet understood. For example, many common diseases—diabetes, high blood pressure, and depression, among them—are known to be influenced by many genes, no one of which contributes to causality in more than a small percentage of cases. It is likely that disorders such as these, which in the past were considered single illnesses, will turn out to be groups of disorders with many different genetic causes.

While knowledge about genetics accumulated throughout the twentieth century, as researchers developed the tools of molecular biology, the rate accelerated dramatically in the last half of the century. The history of this achievement is fascinating and has several general lessons applicable to the search for causality.

Chromosomes were proposed as the source of heredity in the late nineteenth century. The protein deoxyribonucleic acid (DNA) was discovered by Johann Miescher in 1874, and seventy years later, in 1944, Avery and McCloud identified DNA as the conveyor of heredity. In 1953, James Watson and Francis Crick discovered that DNA consists of two strands that twist around each other in a shape called a "double helix" and that each strand consists of repeated sequences of only four chemicals called nucleotides: cytosine, guanine, adenosine, and thymine. The two strands line up such that the nucleotides adenosine and thymine are always opposite or paired with each other, as are guanine and cytosine. Within a decade it was shown that unique sequences of three of these nucleotides guide the formation of each of the twenty amino acids seen in living organisms. These amino acids are then linked together to construct the proteins that are the basis of all life. For example, the series cytosine ("C"), adenosine ("A"), and guanine ("G") or CAG instructs the cellular machinery to make the amino acid glutamate.

Crick and Watson's model explained the mechanisms of many aspects of heredity that previously had been inferred from experiment and observation. For example, Mendel's model required that each parent contribute one of their paired heredity units (genes) to their offspring. The Watson-Crick model proposed that this occurred when the paired strands of the helix unwind during a process called meiosis so that only one of their paired genes is contained in a single egg or sperm. It is the reunification of these at the time of fertilization that leads to the pairing of genes.

Almost fifty years after the Watson and Crick paper was published, two groups of scientists announced that they had been able to put in correct sequence the three billion nucleotides of the twenty-three human chromosomes, known in their entirety as the human genome. This sequencing has been hailed as one of the great achievements of science because it required the putting together in exact order the many millions of small pieces of the twenty-three chromosomes that the researchers had produced. These small pieces were necessary in order to be of a manageable size that could be sequenced by different machines and groups of individuals working simultaneously. The pride and excitement of the scientists involved and of the scientific and lay communities at large when a preliminary draft of the final sequencing was announced in 2001 was palpable, but several surprising questions quickly emerged. First, the very definition of a gene came into question, for it turned out that the strings of nucleotides that appeared to operate as individual genes and make single proteins sometimes did more than this. For example, some nucleotide strings directed the production of several different proteins, each of a different length but sharing parts of the identical amino acid sequence. While this had been known before the sequencing was completed, the frequency and complexity of the process undermined the simple idea of the gene as the basic unit by which information is passed on to the next generation. Currently, several definitions of the gene have emerged, each of which emphasizes a different aspect of the genetic process: (1) the unit that directs the production of a unique protein, (2) the unit that results in the production of an identifiable characteristic (called a "phenotype") of an organism, and (3) the unit that directs the production of a specific RNA (ribonucleic acid) protein that forms the template from which a specific protein is produced.

This surprising turn of events demonstrates that the continued accumulation of knowledge often overturns or requires modification of what had been universally accepted as natural "truth." The concept of the gene was both simple and elegant, and it was supported by one hundred years of research on plants and animals. Specific genes were identified as the cause of eye color in the fruit fly and the determinant of how many legs an insect has. However, many other characteristics that did not follow such a simple model now appear to have other explanations or at least to have a much more complex genesis. Our understanding of causality is sometimes simplified by discovering the mechanisms through which nature operates, but not always.

Over the past decade, these and many other discoveries have challenged the notion that the portion of the genome that codes for the production of proteins is the sole determinant of heredity. Parts of the nucleotide sequence contain instructions to start or stop making proteins, and these too are subject to variation and control. Other parts of the genome have sequences that seem to have no function. They were previously referred to as "junk DNA," but these "noncoding regions" are now known to regulate the amount of a gene product that is produced, among other actions. Perhaps even more surprising, it is now clear that genes can be *modified* long after reproduction is completed, even in adulthood. These *epigenetic* mechanisms are influenced by environmental events and the availability of certain chemicals at specific times in the organism's lifespan. This is a revolutionary discovery that contradicts the century-old dictum that environmental events do not induce changes in the genome that can be passed on to offspring. The claim that environment could change genes was disparaged throughout the last half of the twentieth century as political ("Lysenkoism," after the Russian communist scientist who proposed it) rather than scientific, but it is now accepted because experimental evidence has demonstrated its occurrence.

Another challenge to long-accepted doctrine has been the discovery during the last decade that small RNA molecules play a role in directing gene function. Thus, DNA is not the only purveyor of heredity, a modification of the Avery-McCloud and Watson-Crick models.

One other aspect of nucleotide sequencing bears mentioning: the linear order of base pairs directs, via messenger RNA, not only the amino acid sequence of proteins but also information about how that protein will fold. This folding is crucial to the function of the protein and reflects the fact that proteins are three-dimensional structures. This "higher" level of information influences not only whether the protein is functional but also the degree or amount of its function. Thus, the genome contains information that is both categorical and dimensional.

The previous discussion has focused on the molecular basis of heredity. An equally fascinating story can be told about the discovery of knowledge at the macroscopic levels of populations and time. Darwin and Wallace began the modern revolution in heredity by claiming that their studies of variation among individual organisms leads inexorably to the conclusion that current life forms developed or evolved over many millions of years. Darwin acknowledged that farmers and breeders had demonstrated over many centuries that selective breeding could result in the modification of the traits of an existing species or, as in the breeding of dogs, dramatically varying offspring. What was radical was that Darwin and Wallace proposed that the wide variety of current life forms had emerged gradually over many millions of years from very simple life forms and that a guiding principle of natural selection, or "survival of the fittest," as one of Darwin's contemporaries, Herbert Spencer, called it, was driving this process. Neither Darwin nor Wallace proposed a mechanism by which this occurred, however.

In the 1950s, Ernst Mayr, building on his own work and that of others, proposed a "grand synthesis" of Mendelian genetics and Darwinism. He proposed that three elements are necessary for evolution—randomness (chance introduction of change at the molecular and chromosomal level, for example, by mutation), variation (these changes result in different biological expression or phenotypes), and selection (greater likelihood of survival in specific environments). Inherent in this model is Mayr's suggestion that these act at the level of the organism, not just at the molecular level. Today, evolution is widely although not universally accepted as one of the great intellectual breakthroughs of all time.

How could such a mechanism have come into place? And how can be it proven that evolution, that is, the development or emergence of new species from existing ones, has indeed taken place over hundreds of millions of years? These are grand causal questions whose answers need to be addressed by any schema that strives to place a framework around causality.

Some facts are incontrovertible. The Earth is many billions of years old, and it has undergone geological change since its origin. Rocks of very different ages contain very different fossils whose age can be estimated. These clearly demonstrate both that life forms have changed dramatically over time and that most of the life forms that have existed during the history of the Earth have become extinct. The fossil record further demonstrates that the oldest life forms are less complex than later ones and that similarities can be found in structure across fossils of very different species. The fossil record also clearly demonstrates that fossils recognizable as similar to *Homo sapiens* have been present for several million years at most and less than one hundred thousand years if one sticks very closely to the skeleton of current humans. Clearly, life forms have changed over time. But what has driven or caused this to happen, and how can that be proven?

At the molecular level, the Watson and Crick model provided a mechanism by which variation could occur, since a change in a single base pair could lead to the production of a different amino acid and therefore a different protein. Subsequent research, particularly the sequencing of the genomes of many organisms, has demonstrated many other potential mechanisms by which variation in the genome has occurred. These include certain pieces of the genome that have duplicated, triplicated, or multiplied many more times. Other parts of the genome are made up of sequences that have flipped direction, and still others have moved *en bloc* from one part of the genome to another. Random change can occur as the result of chance, for example, cosmic radiation striking a DNA molecule, but mechanisms that encourage rapid mutation are also built into the genetic machinery of some organisms, as the example of the influenza virus's propensity for producing genetic variation illustrates.

The third element of the grand synthesis, selective pressure, has been shown to result from changes in the environment. Many variants cannot and do not survive, but once a species is established any random change would remain rare unless the change increased the likelihood of survival into the reproductive age. Environmental change could also introduce circumstances that would decrease the likelihood to survive to the age of reproduction, thus decreasing the likelihood of survival or fitness of an organism. For example, mutations remain a major cause of in utero death in humans, and the increased likelihood of virus spread by more efficient reproduction has characterized the major influenza epidemics of the past century. A nice example of intraspecies change in characteristics over the span of one or two generations is the work of Peter and Mary Grant on the Galapagos Islands. They found that changes in the beak size and structure of Darwin finches paralleled weather changes that in turn had affected the availability of different types and sizes of seeds, their primary food supply.

Other examples of change within species over time have been mentioned in previous chapters. In addition to the dramatic ability of the influenza virus to change into more virulent forms, the examples of the nearly universal ability of bacteria to develop resistance to antibiotic medications and of some cancers to develop resistance to anticancer treatments by developing mechanisms to overcome the drugs demonstrate that natural selection does take place under the pressure of environmental change, in this case the introduction by humans of medications.

However, the great unresolved question is how new species develop. Two competing theories have developed, one of which postulates gradual accumulated change and the other, called punctuated equilibrium, which argues for relatively sudden dramatic change that occurs after many millennia of stability. One can find fossil evidence for both theories, but the evidence is still quite weak. On the one hand, the fossil record to date does not show the gradual changes over time that the gradualist theory requires. On the other hand, the lack of gradual change that has been offered as evidence for the punctuated equilibrium theory might merely reflect an inability to detect gradual change because the fossil record is very sparse. Some skeptics have even used the relative paucity of intermediate forms as evidence that evolution has not been proven to occur. During the past twenty-five years many new fossilized life forms that can be considered intermediates have been discovered. For example, the proposed evolution of birds from dinosaurs was originally predicted on the basis of similarities in bone structure. The discovery of unmistakable evidence of several dinosaur species with feathers has provided some evidence of confirmation. Yet, skeptics reply, one should see many more intermediates if indeed birds evolved from dinosaurs. A final convincing demonstration has not yet taken place, in my opinion, but the continued discovery of more intermediate dinosaur-bird composites suggests that a filled-in record of gradual change *may* be in the offing. It remains quite possible that both the gradualist and punctuated equilibrium theories are correct and that specific dramatic environmental events drive some evolution and that the gradual accumulation of change explains change in other species.

Genetics and heredity thus provide useful illustrations of many of the challenges to approaching causality that have arisen throughout this book. First, solutions that seem the simplest, in this case conceptualizing genes as boxcars lined up on a chromosomal train in a specified pattern, is not necessarily correct. A specific sequence of base pairs may code for a single protein in one tissue or at a specific time in the developmental sequence, but at other times it is part of the instruction to produce a different protein. Similarly, questions that once seemed simple, for example, whether diabetes is caused by the environment or innate genetic mechanisms, have turned out to have complex answers. Many illnesses once considered single entities are now known to be influenced by many genes. For diabetes, depression, autism, Alzheimer disease, and Parkinson disease, many more than ten distinct genes have been identified that contribute to their development. Just how these influence the development of each illness is still being worked out, but it is clear that environment plays a significant role in each as well. Perhaps even more intriguing, in diabetes, this gene-environment interaction need not be permanent.

Thus, some people with a gene predisposing to diabetes who become overweight and develop diabetes can have a resolution of the disease and return to normal blood sugar control if they lose the excess weight. Do they still have the "disease" when the manifestations of it are gone?

Thus, the idea that simplicity is beauty and therefore likely to be right needs to be tempered. As the sequencing of the genome and the still unfolding revolution in understanding of genetic mechanisms demonstrate, many aspects of heredity (nature) do not follow a reductionistic idea of single heredity elements. Mayr's emphasis of the idea that evolution acts on organisms, not just molecules, suggests that the concepts of predisposing, precipitating, and programmatic causality will need to be employed to increase our understanding of evolution. Empirical data sometimes demonstrate that simple answers are correct, but at other times they reveal the converse. Occam's law is often right but also often wrong.

Finally, genetics illustrates as well as any topic discussed in this book how answers to questions about causal mechanisms can change over time as new knowledge is learned. Asking questions that are answerable by current methods is a productive strategy, but there is always the possibility that more information will emerge in the future that alters that explanation. The insights of Hume, Heisenberg, and Gödel apply to biology because absolute certainty must be tempered by the appreciation that future gains in knowledge are likely to explain phenomena at a more basic level or from a different point of view.

At an even broader level, the historical development of ideas and theories in genetics and heredity suggests that it is an error to make absolute distinctions between the empiric and the empathic. Darwin's On the Origin of Species brought together many observations that made his ideas plausible and his theory coherent and comprehensive, but he offered only description. He could not suggest plausible biological mechanisms by which natural selection could operate, nor could he suggest scientific tests to support or refute his ideas. While paleontological data that have accumulated bolster the idea that changes occurring over time have fueled evolution, the discovery of many mechanisms by which change can take place has not yet led to a comprehensive proof of the emergence of new species. The emergence in our lifetime of new organisms such as the HIV virus gives further support to the theory of evolution, but even today it is the weaving together of many different threads of evidence into a coherent narrative that adds the greatest support to Darwinism.

The Three-Facet Model Applied

Genes and heredity can be *predisposing*, as in the example of diabetes, or *precipitating*, as in those diseases inherited in a Mendelian dominant pattern. Many normal functions of living organisms are carried out by networks of genes linked sequentially by the production of one protein that then brings about an event in another gene to increase or decrease activity. These *programmatic* systems likely underlie much of normal biology and, when perturbed, disease. The phrase "natural selection" suggests a purpose for genetics—a mechanism that increases the likelihood that an organism's progeny will survive and therefore increase the frequency of that organism's genes. This reflects *purposive* cause, and even staunch defenders of Darwinism such as Richard Dawkins seem to have difficulty avoiding it. Whether one ultimately rejects or accepts purposive cause as meaningful, thoughtful analyses of this issue by writers such as Michael Ruse in *Darwin and Design* and J. Fodor and M. Piattelli-Palmarini in *What Darwin Got Wrong* can help identify where empirical data are needed to address remaining knowledge gaps.

The Mendelian model remains useful and is an example of *categorical* logic. Earlier, many diseases were believed to be caused by changes in a single gene and to operate in a categorical fashion. However, many gene products can be produced in graded amounts, and many characteristics that are influenced by many genes such as height, skin color, and intellectual ability are dimensionally distributed in the population, as discussed in chapter 4. Many systems under genetic control operate in a nonlinear fashion following the characteristics outlined in chapter 5. For example, the twenty-four-hour circadian system appears to result from the accumulation of gene products and small RNA molecules that influence the operation or nonoperation of nonlinear systems including the production of cortisol, diurnal variations in body temperature, and the sleep/wake cycle.

Empirical science has driven the dramatic accumulation of knowledge of the mechanisms by which heredity exerts control and influence on so much of life. Whatever biological mechanisms are invoked to explain the importance of relationships within families of related individuals (whether genetically or by adoption), I would argue that the power of family, tribal, and ethnic ties lies as much in the meaning of these relationships as in any biological underpinning that might be hypothesized or extrapolated from experiments on nonhuman animals. Overlooking the *empathic* power of such relationships would be a grave error for anyone who wishes to govern, to direct groups of people (whether friends, workers, or relatives), or to learn the lessons of history.

ALZHEIMER DISEASE

In 1901, a fifty-one-year-old woman was brought to the main psychiatric hospital in Frankfurt, Germany, by her husband, who reported several problems. For some months she had been accusing her husband of having an affair, something he adamantly denied. She also seemed very forgetful and was having difficulty finding her way around familiar places. She was admitted to the hospital and diagnosed as having dementia, a syndrome characterized by declines in multiple aspects of cognition. She remained hospitalized until she died five years later, at age fifty-six. During her hospitalization she had disrupted sleep, was physically aggressive, and declined dramatically in her ability to care for herself. An autopsy was performed, and Alois Alzheimer, one of the leading neuroscientists (a term that was invented seventy-five years later) of his time examined her brain tissue under the microscope. He identified two distinct abnormal microscopic structures, today called neuritic plaques and neurofibrillary tangles, scattered throughout her brain. He described these as a "Peculiar Disease of the Cerebral Cortex" when he presented the findings at a scientific meeting at the end of 1906 and concluded that, because she was young, this was likely a rare "pre-senile" form of dementia.

In 1967, the British psychiatrist Martin Roth and colleagues demonstrated that plaques and tangles were the predominant abnormality in the brains of elderly people dying with typical senile dementia and that older individuals with normal cognition had few or no plaques and tangles in their brains when they died. Today, Alzheimer disease is recognized as the most common cause of dementia in late life and is a major public health problem.

Thus far, no single cause of Alzheimer disease has been identified. Many facts are known, and these have fostered several hypothesized causal mechanisms that are widely advocated. About *2 percent* (some estimate more) of cases of Alzheimer disease are attributable to genetic abnormalities that are inherited in an autosomal dominant fashion and located on chromosomes 1, 14, and 21. These are *precipitating* causal genes, because everyone with the gene will develop the disease if they live long enough. Each of these genes is known to be involved in the production or destruction of a protein, the beta amyloid precursor protein, that is located in the cell membrane of every neuron, the brain cells that produce and transmit information.

A number of *predisposing* risk factors have been identified. Most powerful is older age. Indeed, at age ninety, the risk of having Alzheimer disease is 30 to 50 percent. This rate continues to increase so that by age ninety-five, the oldest age for which reliable information is available, the risk of developing Alzheimer disease is 15 to 27 percent per year! Other risk factors include having a family history of dementia, head trauma with loss of consciousness at any time in life, less education, midlife high blood pressure, a past history of depression, and being female. The most powerful and best-studied predisposing genetic risk is a normal gene located on chromosome 19 called *APOE*, which directs the productions of a protein that carries lipid molecules in the bloodstream. This gene comes in three forms, labeled $\varepsilon2$, $\varepsilon3$, and $\varepsilon4$. These are "normal" genetic forms or alleles, but inheriting a single copy of the $\varepsilon4$ allele increases the risk of developing Alzheimer disease approximately threefold. Inheriting a copy of the $\varepsilon2$ allele decreases the risk. Since we inherit one copy of every gene from each parent, there are two alleles of each gene. For people who are $\varepsilon4/\varepsilon4$, the risk of developing Alzheimer disease is twelve-to-fifteen-fold greater than someone who has no $\varepsilon4$. However, individuals who are $\varepsilon4/\varepsilon4$ are not at 100 percent risk, since a number of individuals over the age of one hundred who are $\varepsilon4/\varepsilon4$ do not have Alzheimer disease.

There is good evidence that plaques and tangles form in the brain for ten to fifteen years before the first symptom of the illness, but in many individuals the disease onset can be dated to a period of six months to one year. This is a nonlinear *programmatic* causal pattern, suggesting that the crossing of a threshold of cell death or brain system involvement is required to "trigger" the onset of clinical symptoms.

Empirical studies have found that the plaque lesions consist of fragments of the beta amyloid precursor protein. This fragment is referred to as "beta amyloid." Currently, the most widely held causative theory posits that a particular form of the beta amyloid protein is toxic to brain cells and kills them. It is hypothesized that this causes a "cascade" of cell death since the release of a toxic fragment kills a cell that then releases more fragments and kills more cells and so on, until a significant number of cells die. Eventually, enough cells die to impair brain function. The tangle lesion consists of a different protein, the tau protein. An alternative theory posits the tau protein as the causal agent. However, because plaques form first in many individuals and because the three autosomal dominant genetic forms influence the production and breakdown of the beta amyloid protein, the amyloid hypothesis has the most adherents. Still others hypothesize that there is an even earlier unknown causal event that leads to cell destruction and then to beta amyloid release.

Like most diseases, there is a typical pattern of symptom development in Alzheimer disease, although there are many exceptions. The average case of the disease presenting to a clinical center lasts about ten years, and in many individuals the symptoms progress through three stages, each lasting three years. During the first stage, memory symptoms are the most noticeable problem. During the second stage, problems with language, doing everyday activities such as dressing, and problems accurately perceiving the world develop. During the final three-year stage, physical impairments in walking, swallowing, and continence emerge.

Not surprisingly, the location of the plaque and tangle lesions within the brain mirrors where in the brain the functions that underlie these symptoms are located. The earliest cells to die in Alzheimer disease are in the smell and memory centers, explaining why memory loss is a predominant symptom during the first three years of the illness. The disease spreads slowly over time, and in the middle stage, plaques and tangles become common in the areas of the brain that control speech, doing everyday activities, and visual perception. Thus, empirical studies help explain the pattern of symptom development. The specifics of individual symptoms, for example the fact that memory for new information can be severely impaired but old memories from earlier in life can be recalled readily, illustrates the *programmatic* organization of the brain: memory is not a single brain function but the result of multiple brain structures that are linked and organized into a functional whole. The mechanism by which the disease spreads throughout the brain remains unknown.

Many individuals who have Alzheimer disease are unaware of their problems, while a lack of awareness is less common in other diseases that cause dementia. This suggests that this lack of awareness of the disease is related to the death of specific cells in Alzheimer disease, but this empirical claim cannot be proven; some individuals invoke an *empathic* mechanism of psychological "denial" to explain this lack of awareness, a claim that cannot be disproven. Personally, I cite the low frequency of unawareness in other dementias as evidence against that claim.

A stronger case for empathic causal reasoning rests on the observation that many family members and caregivers of people with Alzheimer disease experience distress, frustration, and social isolation. This is empathically understandable given the declining function and other symptoms that are occurring in the person with the disease; the fatigue, financial stress, and difficulty in maintaining social relationships that are associated with caregiving; and the progressive decline of the person with the illness. While such distress can lead caregivers and family members to ask "why" questions that imply a *purposive* cause, the universality of the disease (it is present in every culture and occurs at a similar age-dependent rate in every geographical region of the earth) supports the primacy of the empirical model and of the predisposing, precipitating, and programmatic levels of analysis for explaining the disease.

Treatment can also be organized along the three-facet model. Understanding the mechanisms of both the predisposing and precipitating causes will likely lead to multiple treatment strategies. Since age is such a strong but unmodifiable risk factor, targeting other *predis-posing* risk factors and lessening or delaying their contribution to risk could delay the onset of the disease. Ron Brookmeyer and Claudia Kawas have estimated that a delay of five years could prevent up to 50 percent of cases. Understanding how the *APOE* genes or the proteins they produce ($\epsilon2$, $\epsilon3$, and $\epsilon4$) exert their influence might also lead to delaying strategies. (Recent evidence suggests that they might do so by influencing metabolism of the amyloid precursor protein.) Under-standing the function of the three *precipitating* genes on chromosomes 1, 14, and 21 has already led to the development of a large number of drugs that potentially change how the beta amyloid precursor protein is metabolized or broken down in the brain and how the beta amyloid fragments are removed from the brain, but thus far they have not led to improvement in randomized clinical trials.

Caregivers can be helped by information and emotional support. This makes sense empathically, but empirical data also show these benefits. In addition, studies such as those carried out by the psy-chologist Linda Teri show that improving caregiver knowledge and emotional state can directly benefit both the caregiver and the per-son with the illness, an illustration of the programmatic nature of the effects of the disease and its treatment. Studies also show that spiritual and religious beliefs can lessen the adverse impact of caring for a person with Alzheimer disease. Thus, both *empathic* and *ecclesi-astic* logic can help caregivers and many of those with the illness, but the success of these interventions likely depends on the individuals' beliefs, social and ethnic background, personal strengths and vulner-abilities, and on their social and economic resources. These are link-ages that *empathic* reasoning is best at identifying.

IDENTIFYING THE MANY CAUSES OF HUMAN CONFLICT AND AGGRESSION: THE METHODS OF THE SOCIAL SCIENCES

Aggression has long been a feature of the human condition and is one of the great challenges of our time. Defined here as the inflic-tion of harm on another, it takes many forms, from assault, war, and genocide to physical abuse, sexual abuse, and bullying. Questions about the origins of aggressive behavior and antidotes to its occurrence have interested many of the great thinkers of the past. At a simplistic level, two contrasting narratives emerge from these writings—humans as innately good and humans as innately aggressive. Jean-Jacques Rousseau (1712–1778), for example, described human beings as being born into a primitive state of peacefulness that is stifled and distorted by the impositions of organized society. Thomas Hobbes (1588–1679), in contrast, saw mankind as innately aggressive and therefore in need of institutions to counter that tendency. In his view, it was the recognition by some groups of this need that led to the formation of government.

The May 18, 2012, issue of the leading American general science journal, *Science*, published twenty-four news, review, and public policy articles examining human conflict from many perspectives. These articles, most of which were written by social scientists from disciplines including sociology, archeology, anthropology, social psychology, economics, and political science, provide a range of examples of how current scholars identify the many causes of human aggression. They will be used as a primary source for examining the methods scholars, particularly social scientists, use to untangle causal relationships of a complex issue and how the three-facet model can provide a framework by which a multiplex set of information can be considered as a whole.

Facet 3: Logics or Methods

Empirical data support the contention that humans have innate tendencies to act in an aggressive fashion toward others. Violence has been identified in the vast majority, if not all, cultures that have been studied, and paleontological evidence suggests that aggressive behavior toward other humans has been present in *Homo sapiens* for tens of thousands of years. In almost all cultures studied, males, especially young males, are more violent than females, and multiple lines of evidence correlate higher levels of testosterone and related hormones to aggressive behavior. The association of puberty with rising testosterone levels in males provides one link among these factors. However, not all males are aggressive, and testosterone levels do not explain the wide variation in levels of aggression and violence in different individuals, settings, or eras.

Evidence from the biological and medical sciences links aggressive behavior to several distinct brain systems that interact with one another. There are also multiple brain and hormone systems that underlie positive human interactions such as caring for others and conflict resolution. The aggression and caring systems interact with each other, as well, to form a complex, disseminated network. While these systems are present in all individuals, their function varies across individuals and by situation. From the point of view of the nervous system, then, the views of Rousseau and Hobbes are both correct: humans are innately caring *and* innately aggressive.

Empirical evidence also demonstrates correlations between a range of events and socioeconomic circumstances and the likelihood of conflict at a group level. Multiple studies demonstrate increased rates of a variety of aggressive actions in association with economic deprivation, "oppression" of one group by another, lower status of women, lack of democracy, the presence of nationalism, and widespread discrimination against minorities. These associations do not prove causality, but the use of counterfactual and statistical methods for eliminating other possible reasons for these correlations and the replication of these findings in multiple settings and epochs strengthen the likelihood of a causal, explanatory relationship.

Empathic reasoning ties many historical, interpersonal, and intrapersonal events to violence. History is ripe with events such as war triggered by one group's perceived feelings of disadvantage, hatred, or envy. Narrative plausibility also supports causative linkages between higher rates of aggression and the economic and social variables listed in the previous paragraph.

Ecclesiastic reasoning underpins many calls to peace and violence. Claims of given "rightness" are linked to divine sources and moral/ physical superiority. Belief systems rooted in pacifism, nonviolent resistance, divine right, constitutionalism, nationalism, formal religion, and spirituality are used to justify acts of violence and acts of peace and are cited by those who accept and promulgate that system as a cause of peaceful and violent acts. For those who share in the beliefs, the truth is undisputed, and its rejection reflects a fundamental difference that can lead to conflict. As noted in chapter 10, critics of religion such as Harris and Hutchens link the violence carried out in the name of formal religion as evidence that there is no supreme being because of the contradiction between professed peacefulness and the acts of physical aggression carried out in the name and under the direction of organized religions. Others, such as Goodenough and Collins, highlight the peaceful positives that have derived and see the violence carried out under the auspices of organized religion as a reflection of humankind's lack of perfection. And of course, many human conflicts have arisen from strongly held beliefs unrelated to religion.

Facet 2: Levels of Analysis (The Four P's)

Many of the biological, social, economic, and situational variables discussed above in the section on empirical logic act at the *predisposing* level by influencing the likelihood that an individual or a group will behave in an aggressive or peaceful manner, including the likelihood that they will seek a nonviolent resolution of a disagreement. Another set of variables that predispose to aggression were described by Jared Diamond in his book *Guns, Germs, and Steel*, in which differences in geography, exposure and vulnerability to infectious disease, and availability of natural resources are highlighted as influencing the likelihood that groups used violent or peaceful methods for securing perceived needs, resolving differences, and surviving or failing.

Personality traits of individuals predispose them to act in an aggressive or peaceful manner. These are characteristics of people that are present to some degree or another in all individuals (that is, are universal) in a graded fashion (present in a measurable fashion that follows the normal distribution or bell curve). Impulsivity and stability are two such universal human attributes that correlate with the occurrence of violent and caring behaviors throughout life. In recent years, evidence has emerged that these personality characteristics can be moderated (lessening impulsivity decreases potentially negative outcomes) or encouraged (increasing openness to experience is associated with a greater likelihood of learning from negative experiences). This offers an example of how identifying predispositions to violence could lead to prevention, diminishing the likelihood of an undesirable casual relationship and encouraging a desirable one.

Specific *precipitating* events are often identified as causes of violent interactions. They range from the use of violence to address a perceived verbal insult to the initiation of World War I by the assassination of Archduke Ferdinand. The causal nature of these events usually depends on a close temporal relationship and the strength of the narrative plausibility linking them. Many of the factors identified above as predisposing can also act as precipitants to acts of violence.

The *programmatic* level of analysis usefully guides the analysis of the causes of violence from multiple vantage points. At the biological level, it is the interactions among the multiple brain and endocrine systems that result in whether a nonviolent or violent outcome occurs. The function of these multiple systems is shaped over the lifespan by the social and psychological factors described above. Just how all these variables act is not yet well understood, and the complexity of these interactions makes it unlikely that there will ever be a single final explanatory model.

Many of the social and economic factors identified in the *Science* issue and described above as predisposing or precipitating—economic deprivation, "oppression" of one group by another, lower status of women, lack of democracy, the presence of nationalism, and widespread discrimination against minorities—interact at a programmatic level and lead to conflict that would not have occurred had a different combination or fewer events taken place. The seemingly sudden emergence of many crimes and conflicts can sometimes be best understood as the result of programmatic interactions among several or many biological, social, environmental, and economic variables.

One factor not discussed previously that can be examined at the programmatic level is group identity. Many of the articles in the *Science* issue identify humans' tendency for "us" versus "them" reasoning as innate and expressed at both the individual and the group level. Numerous empirical experiments demonstrate that this tendency to make distinctions between "similar" and "dissimilar" is established or present very early in human development, and observational data from a variety of disciplines demonstrate that the concepts of "otherness" and "in group" identity are universal. On the other hand, the boundaries of group identity change dramatically, sometimes over short periods of time, strongly implying that group identity is shaped by social and environmental variables that act at the group level, even though the predisposition for group identity is innate.

Group-level violence can be understood *programmatically* as emerging from interactions among the predispositions of humans to act aggressively and form social organizations and a wide range of environmental and situational variables. While short-term groups such as mobs and gangs may draw together individuals with similar socioeconomic, demographic, ethnic, geographic, and shared life experiences and beliefs, the sociologist John Levi Martin presents evidence in his book *Social Structures* that many long-lasting (and therefore successful) social and political organizations, such as the U.S. military, developed and are maintained by relationships characterized by the following three elements: inequality among those participating, incompleteness (by which he means there is the flexibility to change and adapt), and lack of the necessity of knowing others personally as a prerequisite for individual success.

The *purposive* level of analysis is prominent in evolutionary models that describe increased likelihood of passing genes to the next generation as the basis for the frequency of many behaviors such as aggressiveness. Darwin identified *group selection* as a mechanism by which gene frequency would increase if survival and the greater reproductive likelihood of individuals were associated with an increased likelihood of survival of the group in which they were members. In relation to human conflict, this model suggests that individuals with traits favoring group identity and aggression would more likely survive in times of deprivation and fighting for scarce resources, and therefore groups of individuals with these traits, such as relatives, would more likely reproduce and increase the frequency of these genes in subsequent generations.

How to explain the existence of traits such as altruism that could lead individuals to sacrifice their lives and therefore not reproduce? In the 1930s, R. A. Fisher (the same Fisher discussed in chapters 4 and 8) and J. B. S. Haldane suggested that actions that increase the reproductive success of one's relatives, even if they cost the life of that individual, could also increase the frequency of that person's genes. Called *kin selection* or *inclusive fitness theory*, this mechanism has been proposed as a genetic basis for behaviors in many species in which individuals do not reproduce but act to increase group survival. Worker bees, for example, play a central role in maintaining the hive but do not mate. The development of a mathematical formulation of kin selection by William Hamilton in the 1960s led to the broad acceptance of this mechanism, but the narrative character of this purposive claim can be seen in the spirited debate that erupted when the biologist E. O. Wilson and colleagues Martin Nowak and Corina Tarnita published a strong rebuttal of kin selection in a 2010 article in the prestigious journal *Nature*. Part of this debate focuses on whether the mathematical formulation of the concept is valid, but those favoring and disputing the concept cite behaviors and roles in a range of different species to bolster their arguments. In one rebuttal letter, 137 scientists framed the debate as follows: "Natural selection [the model favored by Wilson] explains the appearance of design in the living world, and inclusive fitness theory explains *what this design is for* [italics added]." The italicized wording emphasizes the purposive quality of this widely accepted evolutionary mechanism. The two sides agree that evolution has played a major role in the development of social behavior but disagree on whether DNA alone can explain the social roles and behaviors present in many species or whether these are better explained by emergent, programmatic mechanisms. John Levi Martin's model of the emergence of social structures in humans is an example.

I cite this debate among biologists as an example of a disagreement among scientific experts over a causal mechanism. The failure to resolve the question, at least in the present, demonstrates that the use of multiple examples and counterfactuals cannot resolve disagreements if different conclusions are drawn from those examples. Over time, though, additional empirical evidence will likely emerge and convince most or all experts of the accuracy of an explanatory model, whether it is one of these two alternatives or some other.

Purposive causality is also a part of many ecclesiastic systems. Human foibles such as violence are conceptualized in many belief systems as failures to acknowledge or adhere to the values of that system, and purposive causal attribution is applied to the resulting tragedies and disasters. Many belief systems link a failure to accept their basic tenets as a cause of human conflict and aggression. Some identify conflict as a sign of the correctness of their views. Similarly, supporters of evolution see the universal existence of violence and altruism as evidence both of their genetic basis and that they must have survival value and therefore serve the purpose of perpetuating the species. In neither case can a falsifiable empirical study be constructed to disprove the claim, but the ability in both systems to cite many supporting examples of the accuracy of the claim are used to demonstrate its validity.

Facet 1: Models or Logics

Thus far, violence and peace have been discussed as if they are *categorical* states. However, the multiplicity of variables associated with violence and human conflict and the low explanatory power of any one of these variables when considered singly strongly suggests that a categorical causal model of human aggression will rarely if ever be valid. An occasional question might be usefully framed from a categorical perspective, for example, whether World War II would have occurred without Hitler, but this is more a provocation to think through a complex issue than it is a question for which any valid categorical causal attribution can be expected.

Many of the variables identified as empirical predispositions to violence—personality traits, economic deprivation, oppression of one group by another, lower status of women, lack of democracy, the presence of nationalism, and widespread discrimination against minorities—have a *probabilistic* relationship with violence. However, *emergent* causal reasoning provides an approach to combining many of these predisposing and precipitating features into an understanding of why a given conflict or violent act occurred. At the level of the person, a predisposed individual might act in an impulsively violent manner because chance provided several antecedents to occur simultaneously. World War II would not have started when it did had not Pearl Harbor been attacked, but Hitler's increasing aggression in Europe and the United States' oil embargo of Japan in response to what it considered overtly aggressive acts toward other nation-states were predisposing factors that contributed significantly.

Δ Δ Δ

One of the limitations of the *Science* issue is the absence of an integration of these many strands into a comprehensive causal explanation of aggression. Steven Pinker's *The Better Angels of Our Nature: Why Violence Has Declined* is an impressive attempt to do just that. Pinker's approach is twofold. First, he builds upon work first published in the late 1930s by Norbert Elias (1897–1990), a German-born social scientist who moved to England just before the onset of World War II. Pulling together evidence from multiple historical sources, Elias found that rates of violence had been decreasing since medieval times in parallel with the increasing size and organization of nation-states. He attributed this pattern of declining rates of violence to a "civilizing process" fostered by the development of these large governing entities and a parallel increase in the importance and geographic range of commerce. To test Elias's hypothesis, Pinker significantly broadened the range of violent behavior examined and compiled data from an extraordinary range of sources. He confirms that by many measures violence has declined significantly and rather steadily over the past five hundred to six hundred years.

What Pinker does so masterfully, however, is to weave together this evidence that the rate of human conflict has declined over time with numerous historical accounts, experimental studies, and a description of the brain systems involved in violence and self-control into a coherent, comprehensive, and compelling account of human aggression. While much of the evidence he cites uses hypothesis testing, counterfactual reasoning, falsification, and statistical methods for eliminating possible confounds, it is his skillful use of the *narrative* method in linking these many strands of information that convinces the reader (at least this reader) of the plausibility, accuracy, and relative completeness of his attempt to explain the human tendencies toward violence and peacefulness. In doing so, Pinker more than meets Richard Evans's criteria of action (violent behavior in its many forms), happening (historical events in which violence was prominent and data from many groups and nation-states), character (the perpetrators and victims, ranging from individuals to cultures and nation-states) and setting (individuals, groups, living situations, multiple settings, governmental units, compacts, nations). While I found myself disagreeing with some of the conclusions Pinker draws from individual studies or data sets, for example his concurrence with the hypothesis that postpartum blues is attributable to "the emotional implementation of the decision for keeping a child," and believe he neglected to discuss the many epics of the Eastern world that would be less supportive of his hypothesis, the book weaves together so many supporting pieces of evidence that it makes a strong, albeit not irrefutable, claim. It more than meets Bradford Hill's criteria for supporting causal relationships.

Pinker spends two chapters, 114 pages, integrating what is known about the human brain systems involved in aggression ("inner demons") and peacefulness ("better angels") with an extensive review of experiments linking these brain systems to human behaviors that are violent, peaceful, or an intermediate form of these extremes. The challenge he faces in doing so is reflected in his use of modifiers and descriptors such as "plausible" (p. 512), "metaphor" (p. 517), "can be understood" (p. 523), "consistent with" (p. 541), "might" (p. 578), "supports" (p. 606), "could" (p. 614), "in the psychology lab" (p. 681), and "cannot prove" (p. 690) when he links experimental results with real-world occurrences. Pinker's and Elias's conclusion that violence is not inevitable is supported by this combination of historical evidence that violent behavior has declined significantly over many centuries, experimental evidence that the implicated anatomical/physiological systems act in a graded function and therefore are amenable to modulation, and experimental evidence that attitudes and behaviors related to aggression can be influenced by expectations, chance, and environmental factors categorized as sociological, political, and economic.

Since Pinker acknowledges that the causal model he builds is provisional, should he have excluded it from the book and only described the evidence that he believes is important in constructing a final model? Absolutely not, in my opinion. His description of the tentative nature of some of the linkages among and implications of his final model is an acknowledgment that those claims require further study, whether that means more experiments, further gathering of historical data, or more study of historical examples. It is this type of causal modeling that spurs the development of testable hypotheses, experimental studies, and historical reviews, and in doing so moves the search for causes forward.

Δ Δ Δ

Can a general answer be given to the question, "What causes human conflict?" I believe the answer is yes. The human brain has multiple systems and genetically programmed changes over the lifespan that predispose to both violence and peacefulness. Individuals have innate tendencies (temperaments) shaped both by biology and experience that influence their likelihood of acting violently or peacefully, and humans share tendencies to see some others as like or different and to search for causes and purposes. Humans are also programmed to act in groups, and this translates individual tendencies into group actions. Events in the lives of individuals, groups, and nations strongly influence the likelihood of peaceful or violent behaviors. Some can be predicted with reasonable accuracy and their consequences prevented, encouraged, or moderated, but others occur by chance. Human traits and institutions have emerged over millennia that have lessened the frequency and devastation of human violence at the level of the individual, group, and nation-state.

This general description, however, does not answer why specific violent events occur and is limited in its ability to predict future violence and peace. Therefore, narrative methods will always be needed to link the details of individual lives, environments, economic circumstances, beliefs, racial and ethnic backgrounds, predisposing biology, innate differences, chance events, and uncontrollable environmental events. It is the weaving together of the biological, historical, and experimental data into a coherent, comprehensive web on which this important enterprise rests. Thus, understanding the causes of human conflict will continue to require the use of both *narrative* and *empirical* methods, the building of *probabilistic* and *emergent* models to explain it, and the identification of factors that act as *predisposing*, *precipitating*, and *programmatic* elements.

Δ Δ Δ

Do the social sciences still have a role in disentangling causality in complex issues such as human violence and conflict? While it is no more possible to identify the boundaries of "social science" than to define the word "science," the social sciences generally address questions about human interactions and institutions. As Vico suggested more than 250 years ago, the breadth of the questions raised by the social sciences often requires more reliance on narrative/empathic logic and more reliance on top-down reasoning than studies in the physical sciences.

The most dramatic change in the social sciences since Max Weber, one of the founders of sociology and the social sciences in general, is a greater reliance on empirical studies, both experimental and descriptive, and the use of methods such as counterfactual study design, game theory, cost/benefit modeling, and brain imaging—and, as Pinker's discussion of human conflict demonstrates, more effort to integrate data from the biological and physical sciences.

Does this mean that it is time to drop the distinction between the physical/natural sciences and the social sciences, especially since, as noted earlier, narrative reasoning is employed in the biological and physical sciences and falsifiability is sometimes not possible in either the social or the physical sciences? I believe the distinction still has value and suggest that Weber and Jaspers had it right a century ago when they concluded that the methodological differences between the narrative/empathic and empirical are worth acknowledging and noting and that the social sciences will remain more reliant on narrative reasoning. The benefit of maintaining the distinction can be seen in Pinker's final chapter as he attempts to integrate the many lines of evidence he developed throughout the book into an explanation for the declining rate of violence over time and for proposals to continue this trend. His greater reliance on narrative reasoning and less ability to provide empirical support for these conclusions should serve as a source of caution. This is not meant as a dismissal or refutation of his final conclusions but rather as a basis for distinguishing between the empirical data he so deftly weaves together and the conclusions he draws for future action.

A comprehensive theory of human aggression is an admirable goal, and I believe Pinker's book shows that significant progress has been made in constructing one. While a final comprehensive explanation may never be achieved, the social sciences provide frameworks on which many "big topics" that relate to human behavior such as "group identity," "empathy," and "civilizing process" can be studied and integrated. This integration requires narrative/empathic skills that are not necessary in the physical sciences, a reason to keep the broad distinction between the two.

There are, however, limits to the methods used by social scientists that should be acknowledged and are well illustrated in *Better Angels.* They include a reliance on cross-sectional studies from which causal conclusions cannot be drawn but which tempt many to do so; the use of unrepresentative samples such as American college students in experimental studies and drawing conclusions from the results that are applied universally to humans; the belief that the use of technologically advanced techniques such as functional brain MRI allows for conclusions about the direction of causality when the studies are only cross sectional and correlational; the interpretation of experimental findings in a categorical fashion when the results demonstrate correlations that are graded rather than absolute (such as males versus females, high versus low testosterone); and the presentation of values as facts, a particularly difficult issue since one person or group's fact is another person or group's value. These limitations also apply to the physical sciences and will always bedevil studies of big questions such as the origin of the universe, the basis of evolution, and the causes of human aggression. The identification of the limits and strengths of every study, method, or conclusion is the best way to combat undue generalizations and conclusions in both the social and physical sciences.

Skepticism about the value of the social sciences abounds, however. One such attack is known as the Sokal hoax, named for the author who submitted and had published a nonsensical article entitled "Transgressing the Boundaries: Towards a Transformative Hermeneutics of Quantum Gravity" in the journal *Social Text.* Sokal, a physicist at New York University, was targeting postmodernism, an approach that I also find lacking in utility. However, the derisive tone in Sokal's description of the fraud comes across more broadly as an attack on the narrative method and those who employ it. No such dismissive response was directed toward the scientific method in the past decade following multiple revelations of fraudulent articles based on experiments that were never carried out, purposefully manipulated, or grossly misinterpreted—and then published in the first-line journals *Nature*, *Science*, and *Cell*. These scientific frauds were often uncovered by students working in the laboratory of the perpetrator, not by the reviewers (usually senior scientists in the same field), editors, or publishers of the journals that were duped. The point here is that the perpetration of these frauds, whether in the physical or social sciences, is more a comment on the tendency of individuals to act fraudulently than they are indictments of entire fields or their methods.

The past three hundred years has taught us that new methods and different kinds of knowledge will continue to be discovered and incorporated into better validated causal explanations. New and unforeseen mechanisms such as the postnatal epigenetic alteration of the genome and new methods such as the use of counterfactuals will emerge and help explain observations and examine hypotheses that seemed beyond empiricism prior to their discovery. However, the assumptions made at the beginning of the book, that causes exist and that time is directional, may be beyond empiricism or at least not amenable to empirical, experimental proof. The distinctions between the physical/natural sciences and the social sciences may further erode in the future, but for the present both provide answers to important questions. There will always be a range of expertise in applying them, so the skills behind their use should always be valued. Most importantly, it is no longer necessary to avoid the attempt to explain complex phenomena, as Galileo rightly concluded was necessary four hundred years ago.

DEPRESSION

In the introduction I noted that I am often asked about the cause or causes of depression. The most straightforward answer to this question is that we don't know much. While an extensive search for the biological and social roots of depression and bipolar disorder has gone on for fifty years, the information that has been learned has yet to be translated into an explanatory model that can be tested and confirmed or falsified. There are many leads, and much more is known now than half a century ago, but experts cannot even agree about the definition of the term "depression," much less about its about experiential, biological, and social causes. As a result of this lack of consensus, there are disagreements about the interpretation of research findings; the efficacy of psychotherapy, medication, and electroconvulsive therapy (ECT or "shock treatment"); and about the relationships among such widely used constructs as happiness, bipolar disorder, grief, demoralization, discouragement, and clinical or major depression.

As a clinician who has been in practice for thirty-five years, I find the evidence convincing that the word "depression" has multiple meanings. However, many scholars would disagree with this contention, and the argument will be settled only if a biology of several or all of these hypothesized mood states is determined and they are shown to be relatively distinct or essentially similar.

For the purposes of this discussion, I will examine four meanings of the term depression and present the evidence that different causal mechanisms underlie at least these four potential "types" of depression. Knowledge is so limited and disagreement so rampant among experts, though, that this can only be viewed as a proposal that requires further study.

Sadness and Demoralization as Universal Human Experience

The most common conceptualization of depression is that it is a single, universal emotional state in which people differ only in "amount" or severity. This meaning of depression captures what is a universal human experience: the sadness, unhappiness, and low mood that occur in the face of disappointment or an unwanted event or outcome. For most people, this is a transient feeling state that lasts hours or days and does not interfere with daily functioning. Although this is a normal and universally experienced human emotion, it occasionally persists to the point that the person is unable to see that there are ways out of the feeling state. I will refer to this state of persistent sadness by the term "demoralization." This word was first suggested by Jerome Frank, one of my teachers, who emphasized an "inability to see light at the end of the tunnel" as the central feature of the most common reason people seek professional help, whether that help is from a licensed clinician, religious practitioner, or healer. Demoralization may be accompanied by difficulty falling asleep, modest appetite loss, and difficulty engaging in usual activities, but the person does not experience most of the other symptoms of major depression described below.

Facet 2: Levels of Analysis

Because sadness is universal, there must be *predisposing* brain systems that underlie the emotional response to events that are perceived, experienced, or expected to be adverse. Evidence from experimental psychology, neuroimaging studies of normal individuals, and studies of brain-injured individuals who have become depressed suggests that multiple interacting systems underlie the experience of sad and happy mood. The degree or strength of the emotional response to an event is influenced not only by the perceived severity of the stressor but also by other predisposing and modulating factors such as other recent life events, early life experiences, prior experiences of similar events, the availability of social supports, the person's temperament, social expectations of the group and culture in which the person functions, and other environmental factors. The effectiveness of many forms of psychotherapy, best documented for those with a learning component (cognitive behavioral therapy and interpersonal therapy), suggests but does not prove that the risk of becoming demoralized is increased if a person has learned specific "self-defeating" patterns of thinking. Lack of evidence of response to antidepressant medications supports the contention that this form of depression is different from major depression, but this, too, is controversial.

A *precipitating* event is required in this form of depression since the sad mood is defined as the emotional response to a perceived negative event or set of events. People vary in what they consider a negative event and how they assess the severity of this stressor, and the person's use of *narrative* logic significantly influences both.

Given the multiple brain systems that modulate both the expression and experience of sadness and happiness as well as the influence of variables ranging from early life experience to type and severity of the event and the perception of that event by the individual, an integrative, *programmatic* level of analysis will ultimately be needed to understand everyday emotional experience. My hypothesis is that demoralization will be shown to be similar to the state of sadness experienced universally but that usual recovery mechanisms have not returned the person's mood to its baseline state (itself very different from person to person).

Facet 1: Models of Analysis

The universality of sadness as a response to disappointment, the wide differences among people in what triggers demoralization, and the wide range of symptom severity in response to a given stressor argue against a *categorical* causal model of demoralization and in favor of a *probabilistically* operative causal model. Probabilistic algorithms have been developed that summate usual responses to stressors. For example, loss of a home or a hoped-for relationship is much more likely to elicit a high level of demoralization that the loss of a pen or a poor grade on a single test, but there are wide differences among individuals in how they rate disappointment, in temperamental traits such as resilience, and in the availability of buffering social supports. These all affect the probability of becoming demoralized.

Grief: A Universal Response to a Significant Loss

Grief is a universal human experience in which a loss is followed by a relatively brief period of numbness and then by a period of deep emotional and physical hurt that comes and goes but in which the profound feelings of loss/sadness are triggered by a remembrance of the deceased individual or lost object. Finally comes a period of resolution, which can last many months, during which most of the symptoms gradually abate.

The initial period of numbness usually lasts hours or days. People may be surprised that they are not more upset or grief stricken, and they are able to accomplish activities such as notifying people of the death and arranging a culturally appropriate funeral and ceremony/life celebration.

The second "stage" often lasts for several months and is characterized by a range of feelings that include anger, sadness, and emotional turmoil. It is the sudden drop into the depths of sadness and the moving out of this state, sometimes equally rapidly, that characterizes this emotional experience, and this is one of the distinguishing features of the depression of grief. While deep feelings of sadness may persist for days, this fluctuation in emotions is often superimposed. Difficulty falling asleep is common, especially at the beginning of this stage. Appetite may diminish and lead to weight loss of a few pounds, and ideas of blame that have some element of truth (for example, "If I had insisted that my father see the doctor when he first told me he was having stomach pain, maybe they would have caught the cancer earlier") are common. Some individuals are less interested in usually enjoyed activities, but in between episodes of upset most individuals report that their level of pleasure and self-confidence is normal.

The third period is less well defined temporally and symptomatically. The feelings of distress and the physical or vegetative symptoms of difficulty falling asleep and diminished energy and activity are usually gone, but the sudden upset when reminded of the deceased can still occur and in fact may occur for many years. There is, therefore, no defined "end" of grief. For many, though, the feelings of sadness are fully gone after six months or one year, at which point many describe returning to their typical lifelong mood, energy, and outlook on life.

What distinguishes the depression of grief from other forms of sadness, then, is this characteristic course (the sequence of numbness, feelings of turmoil including sadness, and resolution over many months) and the alternation between sudden, deep sadness when the person is reminded of the deceased and periods of relatively normal mood. The depression of demoralization does not have this course or dramatic fluctuations, even though individuals can be demoralized for weeks or months. In major depression, the fluctuations of mood follow a more regular, stereotyped, and pre-dictable pattern, one in which mood is worse at the same time every day. The mood states that accompany a personality disorder may be rapidly fluctuating but are lifelong, whereas the fluctuating mood states of grief are triggered by a loss and are not present most of the person's life.

Grief is universal, but its expression is significantly shaped by cul-ture. For example, among the Navajo, the public expression of emo-tion is very rare and is widely viewed as unacceptable, whereas in many Islamic cultures the public display of grief is understood as a sign of the importance of the relationship with the deceased.

Facet 2: Levels of Analysis

The universal nature of grief and the relatively sequenced pat-tern of its course suggest it is a "hard-wired," brain-based aspect of life and that humans are universally *predisposed* to experiencing it. The loss is the *precipitating* event. The sequential, stage-like course suggests that there is an innate, *programmatic* aspect that directs the emotional experience. At a *purposive* level of analysis, grief is a shared experience and may have been genetically selected for because it fostered or was an outgrowth of group togetherness at the time when task sharing during farming and hunting became beneficial for survival. It may also be a lifelong manifestation of the emotional linkages that underlie childrearing and caring for ones' relatives, and in this fashion is an outgrowth of whatever underlies kin selection theory.

Facet 1: Model of Analysis

This description conceptualizes grief as a *categorical* state because of its stereotyped pattern and linkage to a specific set of precipitants such as the death or loss of a loved one. The existence of expected patterns within a culture but wide variation between cultures sup-ports the notion that the essential nature is inherent or built in and therefore categorically unique but that rituals emerge to support those experiencing it.

Facet 3: Logics (The Three E's)

While *empirical* evidence supports the universality and significant cultural shaping of grief, understanding the meaning of loss uses *empathic/narrative* logic. The strong shaping of the experience of grief by cultural norms demonstrates the importance of grief to the human experience and the central nature of relationships with other humans to our lives.

Major Depression

Clinical or major depression has afflicted humankind for thousands of years. The ancient Greeks referred to it as "melancholia," and Hippocrates' 2,500-year-old description is indistinguishable from the disorder as it is experienced today. There are three central elements of major depression: a change in mood, a change in self-confidence, and a set of changes in bodily function.

The mood change is described variously as sadness, being blue, or feeling low, but about one-third of people do not experience this mood change as sadness and are sometimes adamant, therefore, that they are "not depressed." This is well captured in this quote from William Styron's autobiographical account of his major depression entitled *Darkness Visible*, a title taken from a classic seventeenth-century book by Robert Burton entitled *The Anatomy of Melancholy*: "When I was first aware that I was laid low by the disease, I felt a need, among other things, to register a strong protest against the word 'depression.'" What is clear from Styron's description is that he was experiencing something that "laid him low," that is, took him over and was distinct from any usually experienced state. This captures one of the two elements that distinguish this feeling from the sadness of demoralization. The first is that people with major depression often feel engulfed or captured by a persisting and abiding change in emotional state, which they recognize as distinct from the usual experience of mood (I speculate that this is the reason many individuals experience and report that it is different than the sadness of everyday life). Second, major depression is accompanied by a diminished or lost ability to enjoy and get pleasure from usually enjoyed activities such as reading, talking with friends, participating in athletic activities, or seeing grandchildren. The altered feeling state of major depression is present much, if not all, of every day, is often strikingly worse at the same time every day (referred to as "diurnal variation"), and is relatively unaffected by external events, whether these are positive or negative.

The second feature of major depression is a lowering of the person's usual sense of self-confidence or self-worth. Importantly, this is a change from what is usual for that person and is generally noted by the person. Often, the person attributes the change to a reason or event, making it seem plausible, but it is the presence of this diminution in experienced self-worth that is the central issue.

The third element of major depression consists of a set of physical symptoms that include low energy, poor concentration, loss of appetite and weight, and impaired sleep, in which the ability to fall asleep is maintained but insomnia in the middle of the night and awakening earlier and less refreshed than usual occur nightly. Many people describe feeling "different" in some physical way compared to their usual self, a feeling that is hard for many people to be more specific about and that seems to be what Styron was saying.

Major depression is distinguished from demoralization by this characteristic triad of change in mood, lowered self-confidence, and vegetative, physical symptoms and by its persistence (currently the convention is to require a length of at least two weeks, but a usual episode lasts many months if not treated). In demoralization, the sleep problems more commonly involve difficulty falling asleep rather than early awakening, and the person does not lose the ability to obtain pleasure from usually enjoyed activities. Major depression is distinguished from grief by grief's irregular fluctuations in emotion and sequential pattern over months of numbness, emotional turmoil, and resolution. It is distinguished from depressive personality by the lifelong character of depressive personality versus the periodicity (episodes of the depressive triad that punctuate longer periods of usual or "normal" mood) of major depression.

Stressors are sometimes identified as preceding an episode of major depression, although it can be difficult to determine whether the "precipitant" was in fact a result of the early symptoms, for example, low energy leading to a missed appointment that has adverse consequences. Nonetheless, studies suggest a weak but positive relationship between stress and the development of major depression. Since a stressor or presumed precipitant may or may not be present in major depression but is always present in demoralization, the *presence* of a stressor is not useful in classifying a person's depression as demoralization or major depression, but the *lack of* a precipitant makes major depression more likely.

Facet 1: Models

The existence of a characteristic triad of symptoms that is similar across many different cultures, that occurs for a limited period of time during a person's life, and has been stable for over 2,500 years supports the contention that this is a *categorical* condition and that its cause or causes will be related to a categorical change in the mood-control system. Because our knowledge of this system is rudimentary, there is no basis on which to speculate what the specific changes are and how they will differ from those that might underlie other mood states, but this formulation suggests that it will be a qualitative change in function rather than a change in degree, as is predicted in demoralization.

Studies of twins suggest that 50 to 60 percent of the risk of major depression is genetic: the risk of both twins having the disorder is 50 to 60 percent if they are monozygotic or identical, while it is approximately 10 to 15 percent if they are dizygotic or "fraternal," the same likelihood as one sibling having major depression if another sibling or parent has it. No single genetic abnormality has been identified that explains this risk, but a number of different genes have been implicated. Whether these different genetic risk factors ultimately lead to a single causal pathway or induce depression by multiple mechanisms is unknown but under intensive study.

Depression is often said to be "biochemical" because antidepressant medications are more effective than placebo in treating moderate and severe major depression. It is worth noting, though, that the combination of medication plus psychotherapy is more effective than medication alone. Since the drugs that are effective in treating major depression influence levels of the chemicals serotonin, nor-epinephrine, dopamine, and perhaps acetylcholine, a long-explored theory is that abnormalities in one or several of the brain systems that utilize these neurotransmitters is impaired. However, no consistent change has been found in people with depression. Antidepressant medications influence levels of a number of other brain chemicals, and some induce other changes, including the formation of new brain cells, so it is possible that the mechanism by which they improve depression is yet to be discovered. A counterfactual finding is that drugs that sedate and decrease anxiety, such as barbiturates, alcohol, and benzodiazepines (Valium, Xanax, and Ativan), act on neurotransmitters related to the compound GABA rather than those influenced by antidepressants, a finding suggesting that depression is not a generalized impairment of all brain systems nor a descriptor of all negative emotional states. However, anxiety is a common symptom in major depression, and recent genetic and pharmacologic studies implicate a possible role for the GABA system.

Interestingly, there are differences on functional brain MRI in individuals when they are depressed and not depressed, and these occur in the same or similar brain regions whether the improvement in mood is induced by antidepressant medications or by placebo. This suggests that a categorical change occurs when mood improves and can be interpreted as implying that whatever returns the system to baseline, be it time, medication, psychotherapy, ECT, physical exercise, or deep brain stimulation, leads to a reinstitution of a normally functioning mood system.

As already noted, people who become depressed report more adverse life events in the time period prior to the onset of an episode of major depression than comparison, nondepressed groups. While the direction of causality is in question, it is certainly plausible that genetic or early life predisposing factors make some individuals vulnerable to a categorical disruption of the mood-control system. Likewise, it is plausible that characteristics such as resilience might be protective. Evidence for these suppositions is lacking.

Multiple lines of evidence lead me to speculate that an *emergent* model will be found to explain the categorical appearance of major depression. The differences between major depression and demoralization are not primarily changes in degree—the decline in the ability to experience pleasure and the diminution of self-confidence, the characteristic sleep changes and the diurnal character of the mood changes are not seen in demoralization. Given the likelihood that the same mood system is involved in both demoralization and major depression (and possibly in grief and temperamental depressive traits, as discussed in the next section), I believe the evidence favors a causal model in which the normal responsivity of the mood system to life events is lost or dramatically (qualitatively) altered in clinical depression, whereas it is changed to a probabilistic degree in demoralization. Clearly, this is speculation, and empirical studies will be needed to test this idea.

Facet 3: Logics (The Three E's)

Many of the data points discussed above were developed using *empirical* methods. The ultimate identification of the causes of depression and a determination of whether any proposed schema such as the one presented here is valid will depend on studies that examine testable and at least partially refutable hypotheses. However, *empathic* logic will always be employed to link precipitants to demoralization and grief and possibly to major depression, since what is a precipitant varies widely among people. Empathic logic may also be utilized to examine the question of the relationship between severity of a stressor and likelihood of becoming demoralized, but it remains a challenge to avoid circular reasoning (an event is a greater stressor because it induces more sadness versus a greater degree of sadness inducing a larger brain change).

Facet 2: Levels of Analysis (the Four P's)

Several of the factors identified above as increasing the risk of developing major depression appear to operate as *predisposing* factors since many individuals with these characteristics or experiences, including genetic risk factors, adverse early life circumstances and recent stressors, do not develop major depression.

Many medications, hormones, and drugs of abuse (including steroids, testosterone, the antihypertensive medications reserpine and alphamethyldopa, alcohol, and cocaine) can precipitate major depression. Some medical conditions are associated with increased risk of major depression, and in many of these there is no relationship between the severity of the disease and the incidence of depression. These include brain diseases such as Parkinson disease and multiple sclerosis, cardiovascular disease, and diabetes mellitus. Depression is associated with pancreatic cancer and certain lung cancers, and the depression is sometimes noted *before* the physical symptoms of the underlying disease become noticed and the disease diagnosed. This supports the notion that these diseases induce depression through a hormonal or other chemical messenger mechanism.

Depression as a Personality Characteristic

The final use of the term depression refers to a person's usual or lifelong emotional state. The likelihood that a person will make optimistic or pessimistic interpretations of a situation; will react to a stressor in an unemotional, controlled fashion or in an emotionally expressive, dramatic fashion; and will react to change in a guarded or embracing fashion are universal characteristics shared by all humans, and these become stable, characteristic aspects of each individual by the mid-teen years. The lifelong tendency to be unhappy, then, can be conceptualized as an essential aspect of a person's being, that is, their personality. In this meaning of the term "depression" each individual can be described as having a place on the continuum from a very high likelihood to experience unhappiness to a very low likelihood to do so. As with all shared universal traits, this likelihood is distributed in the population in a normal or bell-curve fashion such that some people have lifelong persistent tendencies to be pessimistic, see the negative rather than the positive, and describe themselves as experiencing minimal pleasure in response to events perceived by many people as positive and happiness inducing. This meaning of the word depression, then, refers to a lifelong temperamental set of characteristics.

Facet 1: Models

By definition, this is a universal characteristic that follows a *predisposing* model of cause. That is, an individual's "place" on the normal distribution will influence his response to a situation or stressor. Having a depressive temperament increases the likelihood that a person will experience unhappiness or a greater degree of unhappiness to a given stressor but does not absolutely determine the response. A *precipitating* stressor interacts with this preexisting tendency to determine what the mood result will be. Thus, the tendency to demoralization is influenced by temperament in a causal fashion, but the many other variables described under demoralization play an additional role.

Final Thoughts on Depression

It is highly likely, although not fully confirmed or mapped out, that a disseminated web of brain systems underlies what we refer to as mood. This system has multiple nodes that integrate inputs and outputs; modulate feelings of happiness, sadness, worry, anxiety, anger, pleasure, and pain; and link to systems that underlie sleep, energy, initiative, risk assessment, cognition, attention, and concentration. Such a system likely has built-in redundancy and compensating modules that underlie resiliency. If this is the case, the proposed multiple "meanings" of depression may reflect functional dysregulation of various modules and groups of modules (subsystems) of this disseminated network, and the various types of depression will be best understood as arising from the *programmatic* nature of the mood system.

The biology of such a complex system will be a challenge to disentangle. Perhaps that is why attempts to understand its biology have thus far yielded limited results. It seems plausible that such a system will have the following characteristics, identified by Kitano:

I. *System controls*, including *feedforward* controls, in which a given stimulus sets off a series of connected steps that link to systems involved in such disparate actions as pleasure, discomfort, and cognition, and *feedback* mechanisms that maintain the system (temperament/depressive personality) but if disordered no longer allow normal modulation of mood, resulting in major depression

2. *Redundancy*, representation of the same system on both sides of the brain

3. *Structural stability*, that is, multiple pathways that regulate responsiveness (temperament), and increase the likelihood that a person will be able to repeat (learn) responses that maximize adaptive responses in order to function in the face of adverse circumstances (resiliency)

4. *Modular design* that allows the person to function in the face of very upsetting circumstances by limiting the dysfunction (e.g., in grief), includes subsystems that increase the likelihood of a beneficial response (resiliency) and limit the adverse (systemwide) effects and outcomes of depression, and provides for input from other modules that underlie modulating responses such as cognition, stress response, pleasure, reward seeking, and the control of innate drives such as eating, sleeping, and reproduction.

In this proposed programmatic, causal model of depression, major depression would be the result of a multimodule dysfunction in which many modules become minimally or unresponsive to the feedback and feedforward mechanisms and thereby cause dysfunction in an emergent fashion. Depressive personality would reflect the stable setpoints of the multiple systems, and demoralization would reflect a downregulation of a single system or small number of systems that regulate emotional responsiveness but not a categorical dysfunction of multiple, linked systems. Grief would result from a set of modules that are linked in preprogrammed fashion to respond to a specific set of triggers. Major depression, then, would be a categorical/*emergent* dysfunction of multiple systems, depressive personality a lifelong setpoint/equilibrium of a series of modules and subsystems that act in a *predisposing* way, grief the prespecified multimodule response to a significant loss, and demoralization the diminished function of modules regulating emotional responsiveness and resiliency.

Δ Δ Δ

A discussion of the cause or causes of depression may seem premature without a consensus among experts about how to conceptualize depression. However, as P. W. Anderson noted forty years ago, the top-down and bottom-up approaches complement each other. Beginning with normal and abnormal mood states and trying to understand their biological underpinnings is one path to discovering the causal mechanisms of depression, but basic studies of how the brain functions at the molecular, cellular, synaptic, and multicellular levels and how these biological entities are affected by changes in gene function, stress hormone levels, etc. will also be needed. Learning how multiple other systems that likely are involved in depression, such as cognition (itself consisting of multiple interacting modular systems), pain, and learning will be necessary, as well, to understand the myriad factors that underlie the causes and expression of depressed mood. The justification for discussing provisional models before more knowledge is forthcoming on these top-down and bottom-up issues is that proposals such as this one generate testable hypotheses. As noted in chapter 5, the large number of components of the mood system and the significant degree of individual variation might preclude a final, fully explanatory model, and the probabilistic implications of individual variation may never allow confident prediction at the level of the individual.

A second justification for generating possible models of the mood system relates to the narrative/empathic underpinning of the question "Why am I depressed?" The question arises in part because of the incomprehensibility of certain mood states and responses and from the human brain's innate search for cause. There are empirical grounds for a very provisional answer, but, like all narratives, the views of the answerer will significantly influence what is proposed. As a clinician, I feel that answering the question to the best of my ability but presenting it as provisional addresses both the desire of people to know why something is happening while indicating that there is no absolute or single, correct answer. This is the double-edged sword of the innate human tendency to seek causes. Narratives are easy to come by but can be harmful as well as helpful. Narratives will never be equivalent to scientifically generated facts, but narrative causal reasoning provides answers that cannot be accessed by empirical methods, at least for the present. I believe both narrative and empirical approaches will always be necessary when seeking cause at the level of the individual because of the inherent inability to know every fact about a system as complex as mood, a bastardized version of Heisenberg's uncertainty principle and Gödel's incompleteness theorem.

Δ Δ Δ

One of the assumptions of this book is that understanding and acknowledging the methods being used can advance the discussion of a complex topic such as the bases of human aggression. It does not solve the problem of complexity, and admittedly it is complex itself, but it is proposed as a counter to overly simplistic models and as a way to embrace and appreciate the strengths and limitations of very different kinds of information and very different methods of information gathering. It provides a framework that incorporates the work of many scholars who have addressed questions of how knowledge accumulates and proposes a language that incorporates their findings. It is an attempt to be both inclusive and pluralistic but not uncritical or eclectic.

Just labeling various factors as "categorical," "predisposing," and "empirical" is certainly not evidence of utility. What I have tried to show in this chapter is that the identification of correlates and the construction of narratives is easy; demonstrating that they are causal and identifying how they contribute causally is much more challenging. This challenge is compounded when topics as complex and multidetermined as violence and depression are considered. Because factors and descriptors as disparate as testosterone level; discrimination based on physical, social, economic or geographic factors; and group identity can be identified as correlated with levels of violence, a comprehensive or at least sophisticated view that incorporates such a wide conceptual and factual range of characteristics can seem daunting if not impossible to integrate. My hope is that the three-facet model will provide a structure that highlights the strengths of the many intellectual tools needed to explain the "why of things."

References

Introduction

Aristotle. *Physics*. Quotation at book 2, part 3, 194b16.

Burns, P. C., R. C. Ewing, and A. Navrotsky. 2012. "Nuclear Fuel in a Reactor Accident." *Science* 335: 1184–1187. Describes what happens during the meltdown of a nuclear reactor core.

Clery, D. 2011. "Current Designs Address Problems in Fukushima Reactors." *Science* 331: 1506.

Perrow, C. 1984. *Normal Accidents*. New York: Harper Collins.

Wald, M. L. 2004. "In Big Blackout, Hindsight Is Not 20/20." *New York Times* (May 13).

——. 2012. "Combination of Errors Led to Power Loss in San Diego." *New York Times* (May 2).

Chapter 1

Aristotle. 1991. *The Metaphysics*. Trans. J. McMahon. Amherst, N.Y.: Prometheus.

Barrow, J. D. 1998. *Impossibility*. New York: Oxford University Press. Traces the history of science and identifies the limits and strengths of the scientific method.

Butterfield, J., ed. 1999. *The Arguments of Time*. New York: Oxford University Press.

Casti, J. L. 2001. "Formally Speaking." *Nature* 411: 527.

Coveney, P., and R. Highfield. 1990. *The Arrow of Time*. New York: Ballantine.

Ekeland, I. 2006. *The Best of All Possible Worlds*. Chicago: University of Chicago Press. Tackles the ultimate question of cause, "Why is the universe the way it is?" Discusses the irreversibility of time at the macroscopic level.

Gazzaniga, M. S. 1985. *The Social Brain: Discovering the Networks of the Mind*. New York: Basic Books.

Geertz, C. 1983. *Local Knowledge*. New York: Basic Books. A classic discussion of the relativity of knowledge.

Gould, S. J. 1987. *Time's Arrow, Time's Cycle*. Cambridge, Mass.: Harvard University Press.

Le Poidevin, R., and M. MacBeath, eds. 1993. *The Philosophy of Time*. New York: Oxford University Press.

McHugh, P. R., and P. Slavney. 1998. *The Perspectives of Psychiatry*. 2nd ed. Baltimore, Md.: Johns Hopkins University Press.

Notturno, M. A., ed. 1994. *Karl Popper, Knowledge, and the Body-Mind Problem*. New York: Routledge.

Nozick, R. 2001. *Invariances: The Structure of the Objective World*. Cambridge, Mass.: Harvard University Press. Claims that truth is relative to space and time.

Palmer, L., and G. Lynch. 2010. "A Kantian View of Space." *Science* 328: 1487–1488. Reviews the evidence, including Wills et al. (2010), demonstrating that the brain system used by a rat to locate itself in space is an "*a priori* pure form," that is, exists in the brain before the rat uses experience to learn where it is. Supports Kant's claim of preexisting models of cognition, relevant here regarding his notion of causality as an innate construct.

Plotnitsky, A. 1994. *Complementarity*. Durham, N.C.: Duke University Press.

Polyn, S. M., et al. 2005. "Category-Specific Cortical Activity Precedes Retrieval During Memory Search." *Science* 310: 1963–1966. Support for the notion of innate causality.

Popper, K. 2002. *The Logic of Scientific Discovery*. London: Taylor & Francis.

Taylor, C. C. W., R. M. Hare, and J. Barnes. 1999. *Greek Philosophers: Socrates, Plato, Aristotle*. Oxford: Oxford University Press.

Trusted, J. 1991. *Physics and Metaphysics: Theories of Space and Time*. New York: Routledge. A readable history of the concept of time in philosophy and science.

Wills, T. J., et al. 2010. "Development of the Hippocampal Cognitive Map in Preweanling Rats." *Science* 328: 1573–1576. Support for the notion of innate causality.

Vico, G. 1999. *New Science*. 3rd ed. New York: Penguin.

Chapter 2

Achlioptas, D., R. M. D'Souza, and J. Spencer. 2009. "Explosive Percolation in Random Networks." *Science* 323: 1453–1457. Demonstrates that a small addition in the connectivity of edge nodes in a very large system can lead to a sudden change mimicking categorical change.

Aristotle. 1991. *The Metaphysics*. Trans. J. McMahon. Amherst, N.Y.: Prometheus.

Korner, S. 1990. *Kant*. London: Penguin. An accessible introduction.

Kuhn, T. S. 1970. *The Structure of Scientific Revolutions*. 2nd ed. Chicago: University of Chicago Press.

Pearl, J. 2000. *Causality*. Cambridge: Cambridge University Press.

Chapter 3

Dehaene, S., et al. 2008. "Log or Linear? Distinct Intuitions of the Number Scale in Western and Amazonian Indigene Cultures." *Science* 320: 1217–1220.

Murphy, G. L. 2002. *The Big Book of Concepts*. Cambridge, Mass.: The MIT Press. Reviews the evidence that some categories seem to be prewired in the brain (innate) while others are not.

Susser, M. 2001. "Glossary: Causality in Public Health Science." *Journal of Epidemiology and Community Health* 55: 376–378.

Chapter 4

Bernstein, P. L. 1996. *Against the Gods: The Remarkable Story of Risk*. New York: John Wiley.

Blaisdell, A., et al. 2006. "Causal Reasoning in Rats." *Science* 311: 1020–1022.

De Duve, C. 2002. "Mysteries of Life: Is There 'something else'?" *Perspectives in Biology and Medicine* 45: 1–15.

Franklin, J. 2001. *The Science of Conjecture: Evidence and Probability Before Pascal*. Baltimore, Md.: The Johns Hopkins University Press. A readable and informative history of probabilistic reasoning. Corrects earlier claims that it did not exist before the 1650s.

Kaplan, R. 1999. *The Nothing That Is: A Natural History of Zero.* New York: Oxford University Press.

King, G. 1989. *Unifying Political Methodology: The Likelihood Theory of Statistical Inference.* Cambridge: Cambridge University Press. Applies probabilistic reasoning and analysis to questions addressed by political scientists.

Pearl, J. 1988. *Probabilistic Reasoning in Intelligent Systems: Networks of Plausible Inference.* San Francisco: Morgan Kaufmann.

Salsburg, D. 2001. *The Lady Tasting Tea.* New York: W. H. Freeman and Company. An accessible history of statistics.

Seife, C. 2000. *Zero: The Biography of a Dangerous Idea.* New York: Penguin Putnam.

Chapter 5

Achlioptas, D., R. M. D'Souza, and J. Spencer. 2009. "Explosive Percolation in Random Networks." *Science* 323: 1453–1457. Demonstrates that a small addition in the connectivity of edge nodes in a very large system can lead to a sudden change mimicking categorical change.

Alon, U. 2007. *An Introduction to Systems Biology.* Boca Raton, Fla.: Chapman & Hall/CRC.

Anderson, P. W. 1972. "More Is Different." *Science* 177: 393–396.

Bak, P. 1996. *How Nature Works: The Science of Self-Organized Criticality.* New York: Copernicus.

Barabási, A.-L. 2002. *Linked: The New Science of Networks.* Cambridge, Mass.: Perseus.

Berlinski, D. 1995. *A Tour of the Calculus.* New York: Random House.

Buchanan, M. 2002. *Nexus.* New York: Norton.

Camazine, S., et al. 2001. *Self-Organization in Biological Systems.* Princeton, N.J.: Princeton University Press.

de Duve, C. 2005. *Singularities.* New York: Cambridge University Press.

Gavin, A.-C., et al. 2006. "Proteome Survey Reveals Modularity of the Yeast Cell Machinery." *Nature* 440: 631–636. This study demonstrates clustering of the "machinery" within cells that make proteins into 257 unique groupings. This supports the notion that there is clustering of cellular processes into modules or networks.

Johnson, S. 2001. *Emergence.* New York: Scribner.

Kitano, H. ed. 2001. *Foundations of Systems Biology.* Cambridge, Mass.: The MIT Press.

Levin, S. 1999. *Fragile Dominion: Complexity and the Commons.* Reading, Mass.: Perseus.

Miller, J. M., and S. E. Page. 2007. *Complex Adaptive Systems*. Princeton, N.J.: Princeton University Press.

Pikovsky, A., M. Rosenblum, and J. Kurths. 2001. *Synchronization*. Cambridge: Cambridge University Press.

Strogatz, S. 2003. *SYNC*. New York: Hyperion.

Vandermeer, J. H., and D. E. Goldberg. 2003. *Population Ecology*. Princeton, N.J.: Princeton University Press.

Wald, M. L. 2012. "Combination of Errors Led to Power Loss in San Diego." *New York Times* (May 2).

Watts, D. J. 2003. *Six Degrees: The Science of a Connected Age*. New York: Norton.

Chapter 6

Ball, P. 2008. "Quantum All the Way." *Nature* 453: 22–25. Discusses the possible primacy of quantum theory over classical and experiential world views.

Galison, P. 2003. *Einstein's Clock, Poincaré's Maps*. New York: Norton.

Gould, S. J. 1999. *Rocks of Ages*. New York: Ballantine.

Haack, S. 2003. *Defending Science—Within Reason*. Amherst, N.Y.: Prometheus.

Holton, G. 2005. *Victory and Vexation in Science: Einstein, Bohr, Heisenberg, and Others*. Cambridge, Mass.: Harvard University Press.

Oreskes, N. 2001. *Plate Tectonics*. Boulder, Colo.: Westview.

Chapter 7

Bennett, M. R., and J. Hasty. 2008. "Genome Rewired." *Nature* 452: 824–825. Demonstrates how genes can be analyzed as acting in networks and how analyzing them in this way gives different information than analyzing them singly.

Couzin-Frankel, J. 2009. "The Promise of a Cure: 20 Years and Counting." *Science* 324: 1504–1507. A history of the hunt for the gene abnormality in cystic fibrosis and the role that its discovery has played in treatment development.

Finch, C. E., and T. B. L. Kirkwood. 2000. *Chance, Development, and Aging*. New York: Oxford University Press.

Friedland, A. E., et al. 2009. "Synthetic Gene Networks That Count." *Science* 324: 1199–1202. The construction of intracellular clocks that "count" in one direction suggests that biological time can be irreversible.

Gudbjartsson, D. F., et al. 2008. "Many Sequence Variants Affecting Diversity of Adult Human Height." *Nature Genetics* 40: 609–615.

Levin, S. 1999. *Fragile Dominion: Complexity and the Commons.* Reading, Mass.: Perseus. Quotation at 44.

Lowenberg, S. 2009. "Guatemala's Malnutrition Crisis." *Lancet* 374: 187–189. It is not merely lack of calories but rather lack of the right kinds of food that leads to the stunting of growth and height.

McArthur, R. H., and E. O. Wilson. 2001 [1967]. *The Theory of Island Biogeography.* Princeton, N.J.: Princeton University Press.

Nüsslein-Volhard, C. 2006. *Coming to Life: How Genes Drive Development.* Carlesbad, Calif.: Kales. A Nobel Prize–winning scientist describes her work on the biology of embryo development, including the role of chance.

Turner, J. S. 2007. *The Tinkerer's Accomplice.* Cambridge, Mass.: Harvard University Press. Offers a "top-down" approach to biology.

Vandermeer, J. H., and D. E. Goldberg. 2003. *Population Ecology.* Princeton, N.J.: Princeton University Press. Quotation at 3.

Visscher, P. M. 2008. "Sizing up Human Height Variation." *Nature Genetics*: 489–490.

Wardle, D. A., et al. 2004. "Ecological Linkages Between Above Ground and Below Ground Biota." *Science* 304: 1629–1633.

Young, I. M., and J. W. Crawford. 2004. "Interactions and Self Organization in the Soil-Microbe Complex." *Science* 304: 1634–1637.

Chapter 8

Morgan, S. L., and C. Winship. 2007. *Counterfactuals and Causal Inference.* New York: Cambridge University Press.

Perrow, C. 1984. *Normal Accidents.* New York: HarperCollins.

Petroski, H. 1992. *To Engineer Is Human: The Role of Failure in Successful Design.* New York: Vintage. "Every case of failure [is] an opportunity to test hypotheses" (232).

Tenner, E. 1996. *Why Things Bite Back.* New York: Knopf. Quotation at 256.

Wald, M. L. 2004. "In Big Blackout, Hindsight Is Not 20/20." *New York Times* (May 13).

——. 2012. "Combination of Errors Led to Power Loss in San Diego." *New York Times* (May 2).

Chapter 9

Carr, E. H. 1961. *What Is History?* New York: Vintage. Quotation at 113.

Couch, T. D., and P. L. Jakab. 2003. *The Wright Brothers and the Invention of the Aerial Age.* Washington, D.C.: National Geographic.

Evans, R. J. 1997. *In Defense of History*. New York: Norton.

——. 2001. *Lying About Hitler: History, Holocaust, and the David Irving Trial*. New York: Basic Books. Quotation on 241, n. 51. British libel law has been changed in recent years, probably in response to this and similar instances and to the perception that the law was being used to protect individuals with high status from criticism.

Frank, J. D., and J. B. Frank. 1991. *Persuasion and Healing*. 3rd ed. Baltimore, Md.: Johns Hopkins University Press. An accessible discussion of rhetorical methods in the modern era.

Gaddis, J. L. 2002. *How Historians Map the Past*. New York: Oxford University Press.

Gazzaniga, M. S. 1985. *The Social Brain: Discovering the Networks of the Mind*. New York: Basic Books.

Greene, B. 1999. *The Elegant Universe*. New York: Vintage.

Judson, H. F. 2004. *The Great Betrayal: Fraud in Science*. Orlando, Fla.: Harcourt. Demonstrates that reliance on the scientific method and replication alone do not protect against false claims in science. The scientific method thus requires allegiance to an ethic of honesty, as does the narrative method.

Lipstadt, D. 1993. *Denying the Holocaust: The Growing Assault on Truth and Memory*. New York: Plume.

Luria, A. R. 1987. *The Mind of a Mnemonist*. Cambridge, Mass.: Harvard University Press. A fascinating description of a person whose memory was "too good."

McHugh, P. R., and P. S. Slavney. 1998. *The Perspectives of Psychiatry*. 2nd ed. Baltimore, Md.: Johns Hopkins University Press.

Nisbet, R. 1976. "Many Toquevilles." *American Scholar* 46: 59–75.

Novick, P. 1988. *That Noble Dream: The "Objectivity Question" and the American Historical Profession*. Cambridge: Cambridge University Press.

Roberts, R. H., and J. M. M. Good, eds. 1993. *The Recovery of Rhetoric*. Charlottesville: University Press of Virginia.

Simons, H., ed. 1989. *Rhetoric in the Human Sciences*. London: Sage.

Toulmin, S. 2001. *Return to Reason*. Cambridge, Mass.: Harvard University Press. Quotation at 8.

Chapter 10

Armstrong, K. 1994. *A History of God*. New York: Ballantine. Quotation at 253, n. 59.

Brooke, J. H. 1991. *Science and Religion: Some Historical Perspectives*. Cambridge: Cambridge University Press. Quotation at 17.

Burtt, E. A. 1954. *The Metaphysical Foundations of Modern Science*. Garden City, N.Y.: Doubleday. An old but unsurpassed discussion of the role of religion in the lives of great scientists.

Collins, F. 2006. *The Language of God*. Detroit, Mich.: Thomson Gale.

Dawkins, R. 2008. *The God Delusion*. New York: First Mariner.

Dubuisson, D. 2003. *The Western Construction of Religion*. Trans. W. Sayers. Baltimore, Md.: Johns Hopkins University Press.

Franklin, J. 2001. *The Science of Conjecture: Evidence and Probability Before Pascal*. Baltimore, Md.: Johns Hopkins University Press.

Goodenough, U. 1998. *The Sacred Depths of Nature*. New York: Oxford University Press. Quotation at xvi.

Gould, S. J. 1999. *Rocks of Ages: Science and Religion in the Fullness of Life*. New York: Ballantine.

Groopman, J. 2004. *The Anatomy of Hope*. New York: Random House.

Harris, S. 2004. *The End of Faith: Religion, Terror, and the Future of Reason*. New York: Norton.

Holton, G. 2005. *Victory and Vexation in Science: Einstein, Bohr, Heisenberg, and Others*. Cambridge, Mass.: Harvard University Press. The chapters on Einstein and Rabi demonstrate how religious views influenced their work even though they eschewed formal religion.

James, W. 1999. *The Varieties of Religious Experience*. New York: Random House. Quotations at 532, 545. The insights of this classic introduction to the topic remain informative more than one hundred years after it was written.

Larson, E. J., and L. Witham. 1997. "Scientists Are Still Keeping the Faith." *Nature* 386: 435–436.

Maugham, W. S. 1999 [1916]. *Of Human Bondage*. New York: Random House.

Zukav, G. 1979. *The Dancing Wu Li Masters*. New York: Bantam.

Chapter 11

Abbot, P., et al. 2011. "Inclusive Fitness Theory and Eusociality." *Nature* 471: 1057–1060. Quotation at 1057. Rebuts the rejection of kin selection theory by Nowak, Tarnita, and Wilson (2010).

Alexopoulos, G. S. 2005. "Depression in the Elderly." *Lancet* 365, no. 9475: 1961–1970.

Alzheimer, A. 1987. "About a Peculiar Disease of the Cerebral Cortex." Trans. L. Jarvik and H. Greenson. *Alzheimer Diseases and Associated Disorders* 1: 7–8.

Brookmeyer, R., S. Gray, and C. Kawas. 1998. "Projections of Alzheimer's Disease in the United States and the Public Health Impact of Delaying Disease Onset." *American Journal of Public Health* 88: 1337–1342.

Brooks, P. 2000. *Troubling Confessions.* Chicago: University of Chicago Press. Demonstrates that even confessions may not be truthful and that they can be induced and thereby falsely identify causal guilt.

Burton, R. 1977 [1621]. *The Anatomy of Melancholia.* New York: Vintage.

Carey, N. 2012. *The Epigenetics Revolution.* New York: Columbia University Press.

Cramer, P. E., et al. 2012. "ApoE-Directed Therapeutics Rapidly Clear ß-amyloid and Reverse Deficits in AD Mouse Models." *Science* 335: 1503–1510.

Corrada, M. M., et al. 2008. "Prevalence of Dementia After Age 90: Results from the 90+ Study." *Neurology* 71: 337–343.

Dennett, D. C. 1995. *Darwin's Dangerous Idea: Evolution and the Meaning of Life.* New York: Simon and Schuster. A strong proponent of Darwinism lays out his case.

Diamond, J. 1997. *Guns, Germs, and Steel: The Fates of Human Societies.* New York: Norton.

Elias, N. 2000. *The Civilizing Process: Sociogenetic and Psychogenetic Investigations.* Rev. ed. Cambridge, Mass.: Blackwell.

Fodor, J., and M. Piattelli-Palmarini. 2011. *What Darwin Got Wrong.* New York: Picador. Two evolutionary scientists describe some of the limitations of Darwinism.

Harvey, C. D., et al. 2012. "Choice-Specific Sequences in Parietal Cortex During a Virtual-Navigation Decision Task." *Nature* 484: 62–64. Evidence that memory for task performance is the result of the coherent, sequenced firing of sets of neurons rather than being directed by a single neuron. This *programmatic* organization is a possible mechanism for the *emergence* of behavior.

Heng, H. H. Q. 2008. "The Conflict Between Complex Systems and Reductionism." *Journal of the American Medical Association* 300: 1580–1581. Merely recognizing the complexity of a system does not guarantee that the application of this knowledge will solve problems.

Johnson, C. J., et al. 2008. "Structural Insights Into a Circadian Oscillator." *Science* 322: 697–701. Demonstrates how (and why, since they cannot operate in the opposite direction) biological clocks are unidirectional.

Judson, H. F. 2004. *The Great Betrayal: Fraud in Science.* Orlando, Fla.: Harcourt.

Kupfer, D. J., et al. 2012. "Major Depressive Disorder: New Clinical, Neurobiological, and Treatment Perspectives." *Lancet* 379: 1045–1054.

Lyketsos, C., et al. 2008. *Psychiatric Aspects of Neurological Disease*. New York: Oxford.

Mace, N. L., and P. V. Rabins. 2011. *The Thirty-Six-Hour Day*. 5th ed.. Baltimore, Md.: Johns Hopkins University Press.

Martin, J. L. 2009. *Social Structures*. Princeton, N.J.: Princeton University Press.

Mayr, E. 2001. *What Evolution Is*. New York: Basic Books.

Morgan, S. L., and C. Winship. 2007. *Counterfactuals and Causal Inference*. New York: Cambridge University Press.

Nowak, M. A., C. E. Tarnita, and E. O. Wilson. 2010. "The Evolution of Eusociality." *Nature* 466: 1057–1062. Wilson and colleagues' rejection of kin selection theory. Rebutted by Abbot et al. (2011).

Okasha, S. 2006. *Evolution and the Levels of Selection*. New York: Oxford University Press.

Pinker, S. 2011. *The Better Angels of Our Nature: Why Violence Has Declined*. New York: Viking. Quotation at 418.

Posner, R. 2001. *Frontiers of Legal Theory*. Cambridge, Mass.: Harvard University Press.

Rabins, P. V., et al. 1982. "The Impact of Dementia on the Family." *Journal of the American Medical Association* 248: 333–335.

Ruse, M. 2003. *Darwin and Design*. Cambridge, Mass.: Harvard University Press.

Sokal, A. D. 2000. *The Sokal Hoax: The Sham That Shook the Academy*. Lincoln: University of Nebraska Press.

Styron, W. 1992. *Darkness Visible*. New York: Vintage.

Timberg, C., and D. Halperin. 2012. *Tinderbox: How the West Sparked the AIDS Epidemic and How the World Can Finally Overcome It*. New York: Penguin.

Volberding, P. A., and S. G. Deeks. 2010. "Antiretroviral Therapy and Management of HIV Infections." *Lancet* 376: 49–62.

Weiner, J. 1994. *The Beak of the Finch: A Story of Evolution in Our Time*. New York: Knopf. A description of the work of the Grants that demonstrates selection acting over a short period of time.

Index

Abrahamic tradition, 191–93

accidents: complexity and, 155; design error, 154; epidemiology and, 153–56; implications regarding, 153; levels of analysis and, 154; Perrow and, 2, 153–56; single part, 153; subsystems, 154; Tenner and, 156; unit, 153–54

accumulation of change, 66–67

acquired immune deficiency syndrome (AIDS). *See* HIV/AIDS

Against the Gods: The Remarkable Story of Risk (Bernstein), 52

aggression: causal models and, 232–33; ecclesiastic logic and, 227–28; emergent model and, 235–36; empathic method and, 227, 235–36; empirical method and, 226–27, 235–36; overview about, 225–26; Pinker and, 233–35, 236–37; precipitating cause and, 229; predisposing cause

and, 228–29; probabilistic model and, 235–36; programmatic cause and, 229–30; purposive cause and, 230–32

AIDS. *See* HIV/AIDS

airplane invention, 171–72

Alzheimer, Alois, 221

Alzheimer disease: ecclesiastic logic and, 225; empathic method and, 224, 225; empirical method and, 223; historical background about, 221–22; neuritic plaque and, 221–22, 223–24; neurofibrillary tangles and, 221–22, 223–24; precipitating cause and, 222, 224–25; predisposing cause and, 222, 224–25; programmatic cause and, 223, 224, 225; purposive cause and, 224; symptom patterns, 223–24; treatment of, 224–25

Anderson, P. W., 72

APOE gene, 222